IET MATERIALS, CIRCUITS AND DEVICES SERIES 59

Analysis and Design of CMOS Clocking Circuits for Low Phase Noise

Other volumes in this series:

Volume 2	**Analogue IC Design: The current-mode approach** C. Toumazou, F.J. Lidgey and D.G. Haigh (Editors)
Volume 3	**Analogue–Digital ASICs: Circuit techniques, design tools and applications** R.S. Soin, F. Maloberti and J. France (Editors)
Volume 4	**Algorithmic and Knowledge-Based CAD for VLSI** G.E. Taylor and G. Russell (Editors)
Volume 5	**Switched Currents: An analogue technique for digital technology** C. Toumazou, J.B.C. Hughes and N.C. Battersby (Editors)
Volume 6	**High-Frequency Circuit Engineering** F. Nibler *et al.*
Volume 8	**Low-Power High-Frequency Microelectronics: A unified approach** G. Machado (Editor)
Volume 9	**VLSI Testing: Digital and mixed analogue/digital techniques** S.L. Hurst
Volume 10	**Distributed Feedback Semiconductor Lasers** J.E. Carroll, J.E.A. Whiteaway and R.G.S. Plumb
Volume 11	**Selected Topics in Advanced Solid State and Fibre Optic Sensors** S.M. Vaezi-Nejad (Editor)
Volume 12	**Strained Silicon Heterostructures: Materials and devices** C.K. Maiti, N.B. Chakrabarti and S.K. Ray
Volume 13	**RFIC and MMIC Design and Technology** I.D. Robertson and S. Lucyzyn (Editors)
Volume 14	**Design of High Frequency Integrated Analogue Filters** Y. Sun (Editor)
Volume 15	**Foundations of Digital Signal Processing: Theory, algorithms and hardware design** P. Gaydecki
Volume 16	**Wireless Communications Circuits and Systems** Y. Sun (Editor)
Volume 17	**The Switching Function: Analysis of power electronic circuits** C. Marouchos
Volume 18	**System on Chip: Next generation electronics** B. Al-Hashimi (Editor)
Volume 19	**Test and Diagnosis of Analogue, Mixed-Signal and RF Integrated Circuits: The system on chip approach** Y. Sun (Editor)
Volume 20	**Low Power and Low Voltage Circuit Design with the FGMOS Transistor** E. Rodriguez-Villegas
Volume 21	**Technology Computer Aided Design for Si, SiGe and GaAs Integrated Circuits** C.K. Maiti and G.A. Armstrong
Volume 22	**Nanotechnologies** M. Wautelet *et al.*
Volume 23	**Understandable Electric Circuits** M. Wang
Volume 24	**Fundamentals of Electromagnetic Levitation: Engineering sustainability through efficiency** A.J. Sangster
Volume 25	**Optical MEMS for Chemical Analysis and Biomedicine** H. Jiang (Editor)
Volume 26	**High Speed Data Converters** A.M.A. Ali
Volume 27	**Nano-Scaled Semiconductor Devices** E.A. Gutiérrez-D (Editor)
Volume 28	**Security and Privacy for Big Data, Cloud Computing and Applications** L. Wang, W. Ren, K.R. Choo and F. Xhafa (Editors)
Volume 29	**Nano-CMOS and Post-CMOS Electronics: Devices and modelling** Saraju P. Mohanty and Ashok Srivastava
Volume 30	**Nano-CMOS and Post-CMOS Electronics: Circuits and design** Saraju P. Mohanty and Ashok Srivastava
Volume 32	**Oscillator Circuits: Frontiers in design, analysis and applications** Y. Nishio (Editor)
Volume 33	**High Frequency MOSFET Gate Drivers** Z. Zhang and Y. Liu
Volume 34	**RF and Microwave Module Level Design and Integration** M. Almalkawi
Volume 35	**Design of Terahertz CMOS Integrated Circuits for High-Speed Wireless Communication** M. Fujishima and S. Amakawa
Volume 38	**System Design with Memristor Technologies** L. Guckert and E.E. Swartzlander Jr.
Volume 39	**Functionality-Enhanced Devices: An alternative to Moore's law** P.-E. Gaillardon (Editor)
Volume 40	**Digitally Enhanced Mixed Signal Systems** C. Jabbour, P. Desgreys and D. Dallett (Editors)

Volume 43	**Negative Group Delay Devices: From concepts to applications** B. Ravelo (Editor)
Volume 45	**Characterisation and Control of Defects in Semiconductors** F. Tuomisto (Editor)
Volume 47	**Understandable Electric Circuits: Key concepts, 2nd Edition** M. Wang
Volume 48	**Gyrators, Simulated Inductors and Related Immittances: Realizations and applications** R. Senani, D.R. Bhaskar, V.K. Singh, and A.K. Singh
Volume 49	**Advanced Technologies for Next Generation integrated Circuits** A. Srivastava and S. Mohanty (Editors)
Volume 51	**Modelling Methodologies in Analogue Integrated Circuit Design** G. Dundar and M.B. Yelten (Editors)
Volume 53	**VLSI Architectures for Future Video Coding** M. Martina (Editor)
Volume 54	**Advances in High-Power Fiber and Diode Laser Engineering** Ivan Divliansky (Editor)
Volume 55	**Hardware Architectures for Deep Learning** M. Daneshtalab and M. Modarressi
Volume 58	**Magnetorheological Materials and their Applications** S. Choi and W. Li (Editors)
Volume 60	**IP Core Protection and Hardware-Assisted Security for Consumer Electronics** A. Sengupta and S. Mohanty
Volume 64	**Phase-Locked Frequency generation and Clocking: Architectures and circuits for modem wireless and wireline systems** W. Rhee (Editor)
Volume 65	**MEMS Resonator Filters** Rajendra M. Patrikar (Editor)
Volume 67	**Frontiers in Securing IP Cores; Forensic detective control and obfuscation techniques** A Sengupta
Volume 68	**High Quality Liquid Crystal Displays and Smart Devices: Vol. 1 and Vol. 2** S. Ishihara, S. Kobayashi and Y. Ukai (Editors)
Volume 69	**Fibre Bragg Gratings in Harsh and Space Environments: Principles and applications** B. Aïssa, E I. Haddad, R.V. Kruzelecky, and W.R. Jamroz
Volume 70	**Self-Healing Materials: From fundamental concepts to advanced space and electronics applications, 2nd Edition** B. Aïssa, E.I. Haddad, R.V. Kruzelecky, and W.R. Jamroz
Volume 71	**Radio Frequency and Microwave Power Amplifiers: Vol. 1 and Vol. 2** A. Grebennikov (Editor)
Volume 73	**VLSI and Post-CMOS Electronics Volume 1: VLSI and Post-CMOS Electronics and Volume 2: Materials, devices and interconnects** R. Dhiman and R. Chandel (Editors)

Analysis and Design of CMOS Clocking Circuits for Low Phase Noise

Woorham Bae and Deog-Kyoon Jeong

The Institution of Engineering and Technology

Published by The Institution of Engineering and Technology, London, United Kingdom

The Institution of Engineering and Technology is registered as a Charity in England & Wales (no. 211014) and Scotland (no. SC038698).

First published 2020

The Institution of Engineering and Technology
Michael Faraday House
Six Hills Way, Stevenage
Herts, SG1 2AY, United Kingdom

www.theiet.org

British Library Cataloguing in Publication Data
A catalogue record for this product is available from the British Library

ISBN 978-1-78561-801-7 (hardback)
ISBN 978-1-78561-802-4 (PDF)

Typeset in India by MPS Limited

Contents

About the authors

Woorham Bae received the B.S. and Ph.D. degrees in Electrical and Computer Engineering from Seoul National University, Seoul, Korea, in 2010 and 2016, respectively.

In 2016, he was with the Inter-University Semiconductor Research Center, Seoul National University, Seoul, Korea. From 2017 to 2019, he was with the University of California, Berkeley, CA, as a Postdoctoral Researcher. He is currently a Senior SerDes Engineer with Ayar Labs, Santa Clara, CA. His current research interests include integrated circuits for silicon photonics, high-speed I/O circuits and architectures, nonvolatile memory systems, and agile hardware design methodology.

Dr. Bae received the IEEE Circuits and Systems Society Outstanding Young Author Award in 2018, the Distinguished Ph.D. Dissertation Award from the Department of Electrical and Computer Engineering, Seoul National University in 2016, the IEEE Circuits and Systems Society Pre-Doctoral Scholarship in 2016, and the IEEE Solid-State Circuits Society STG Award in 2015.

Deog-Kyoon Jeong received the B.S. and M.S. degrees in Electronics Engineering from Seoul National University, Seoul, South Korea, in 1981 and 1984, respectively, and the Ph.D. degree in Electrical Engineering and Computer Sciences from the University of California at Berkeley, Berkeley, CA, USA, in 1989.

From 1989 to 1991, he was a member of the Technical Staff with Texas Instruments, Dallas, TX, USA. He worked on the modeling and design of BiCMOS gates and the single-chip implementation of the SPARC architecture. Then, he joined the faculty of the Department of Electronics Engineering and Inter-University Semiconductor Research Center, Seoul National University, Seoul, South Korea, where he is currently an Endowed-Chair Professor. He was one of cofounders of Silicon Image, Sunnyvale, CA, now Lattice Semiconductor, which specialized in digital interface circuits for video displays such as DVI and HDMI. His main research interests include the design of high-speed I/O circuits, phase-locked loops, and memory system architecture.

Dr. Jeong was a recipient of the ISSCC Takuo Sugano Award in 2005 for Outstanding Far-East Paper. He is a Fellow of the IEEE.

Preface

For decades, precise clock generation circuit has been playing an important role in various electronics applications, including RF systems, microprocessors, DRAM interfaces, and wireline communications, mainly because achievable performance for such applications highly depends on the precision of the clock. Phase noise and jitter are the performance metrics for evaluating how precise a clock is. This is the reason why understanding of phase noise and jitter is really important, and why system engineers as well as circuit engineers should be aware of them.

As a result, several theories to analyze phase noise and jitter, and innumerable implementations of clocking circuits to suppress phase noise and jitter have been proposed over the past few decades. Nevertheless, development and exploration on this topic are still very active, as electronics is evolving for better performance. Hundreds of technical papers are presented every single year. That means it is getting more difficult for an entry-level engineer to catch up with recent technologies. That is, the engineer should study the history of the decades of developments and researches while keeping up with the state-of-the-art technology. A tragedy here is, textbooks providing a good explanation for classical theories/circuits, however, are too outdated, whereas recent papers present state-of-the-arts but are hard to understand without a sufficient background or experience. The main purpose of this book is bridging the gap between the classical theories and recent innovations.

This book comprises eleven chapters and four appendixes as follows. Chapter 1 provides a brief introduction on how the jitter degrades an electronic system's performance. Chapter 2 reviews a general theory of Fourier transform and power spectral density, and furthermore gives the definition and physical relation of phase noise and jitter. Chapter 3 introduces basic theory and primary implementation of CMOS LC and ring oscillators. Chapter 4 focuses on the phase noise theories for CMOS oscillators. Classical Leeson's model and Hajimiri's linear time-variant (LTV) model are also covered.

Chapter 5 describes basic concept and building blocks of conventional clocking circuits, phase-locked loop (PLL), and delay-locked loop (DLL) which are intended to suppress the phase noise of CMOS oscillator. Chapter 6 deals with the noise sources of PLL and their contribution to the clock jitter. Chapter 7 focuses on the jitter analysis for DLL and compares PLL and DLL.

Chapter 8, 9, 10, and 11 present basic theories and circuit implementations of the state-of-the-art techniques for producing a low phase noise clock. Chapter 8 deals with subsampling PLLs, where reduction of in-band phase noise is persuaded. Chapter 9 describes all-digital PLL/DLLs, which focus on overcoming limitations of analog implementation and taking advantages of CMOS process scaling. Injection-locked oscillators (ILOs) and clock-multiplying DLLs (MDLLs) are presented in Chapter 10 and Chapter 11, respectively.

Even though the authors did their best in the preparation of the manuscript, there will be errors and mistakes because of human imperfections. So, the authors would like to solicit feedback on this book. The authors will be happy to get comments, suggestions, and criticisms via e-mail at the following address.

Woorham Bae: woorham@ayarlabs.com
Deog-Kyoon Jeong: dkjeong@snu.ac.kr

Acknowledgments

The authors would like to thank all those who contributed to this manuscript. The authors would like to acknowledge Olivia Wilkins, Assistant Editor, and Sarah Lynch, Commissioning Editor, of the IET Books for their kind support and assistance on the publication of this book. The authors sincerely appreciate Dr. Sung-Yong Cho (Samsung Electronics), Dr. Han-Gon Ko (Seoul National University), Dr. Min-Seong Choo (NASA Ames Research Center), and Dr. Kwanseo Park (University of California, Berkeley) for their proof-reading and invaluable feedback on the manuscript.

Lastly, I dedicate this book to my wife, Kyung Jean Yoon, for the continual encouragement and support.

On behalf of the authors
Woorham Bae
Santa Clara, CA
April 2020

Chapter 1
Introduction

Electronics has always been a fight against noise. Because an electrical signal is generally expressed in a two-dimensional way, such as voltage (or current) in y-axis and time in x-axis, the noise can also be classified into two primary types, such as voltage noise and timing noise. Figure 1.1 shows a simple example of how the voltage noise distorts an analog signal and a digital signal. For an analog signal, the voltage noise is much straightforward to understand since it is directly related to signal-to-noise ratio (SNR). However, for a digital signal, because of the inherent noise margin of a digital complementary metal–oxide semiconductor (CMOS) circuit, all of the noise components within the noise margin are removed and the digital signal is restored intact. However, the noise injected during the transition of digital signal is converted to the timing noise, instead of being removed.

Figure 1.2 shows an example of an effect of the timing noise on an analog circuit and a digital circuit. In almost every electronic circuits, the timing is defined by a clock signal. The effect of a noisy clock on an analog sample-and-hold circuit and a digital combinational logic is described in Figure 1.2(a) and (b), respectively. In the sample-and-hold circuit, the sampling switch is turned on when the clock is high, and the sampling capacitor stores the sampled voltage value when the sampling switch is turned off. When we assume an ideal switch, the input voltage is sampled every cycle at the exact moment of the clock rising edges. Therefore, the sampled voltage is affected by the timing uncertainty of the clock. The timing noise, loosely defined as the variation of threshold-crossing time in the time-axis is called jitter. If a jitter of Δt is introduced at a certain cycle, the sampling error ΔV is expressed as

$$\Delta V = \frac{d}{dt} V_{in}(t) \cdot \Delta t \tag{1.1}$$

When we assume a sinusoidal input, $A\sin(2\pi ft)$, the maximum sampling error is obtained as

$$\Delta V = A \cdot 2\pi f \cdot \Delta t \tag{1.2}$$

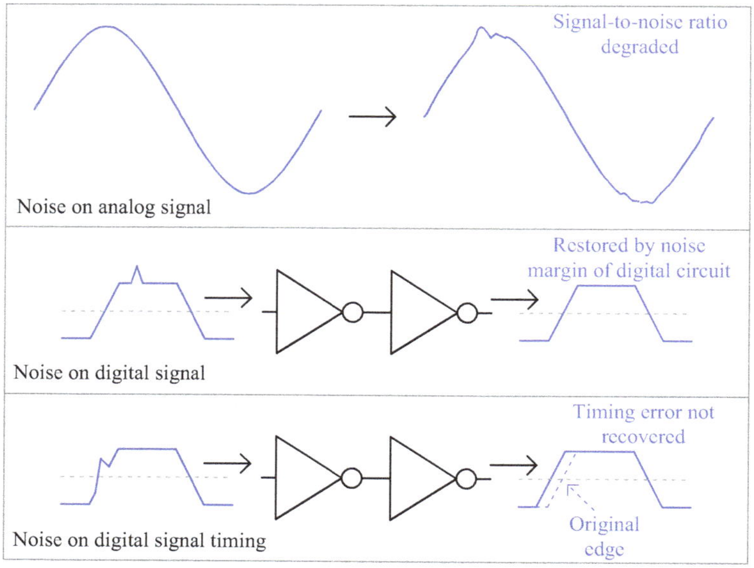

Figure 1.1 Effect of additive noise on analog and digital signals

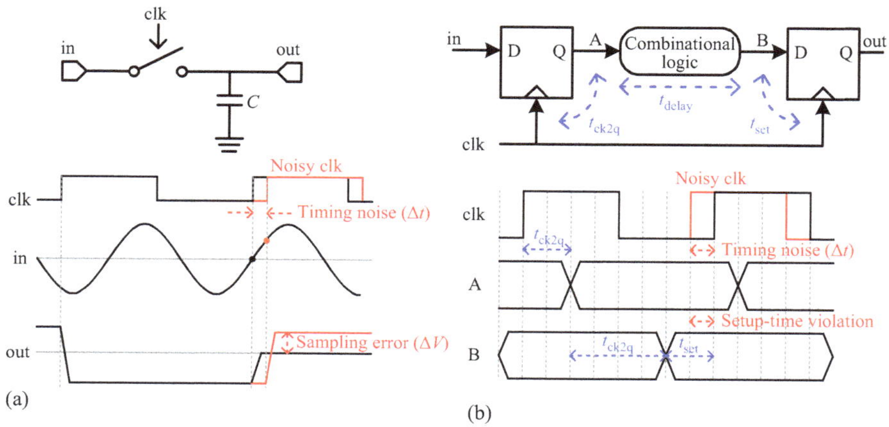

Figure 1.2 Effect of timing noise on (a) sample-and-hold circuit and (b) setup-time violation in digital logic

We can calculate the SNR caused by the timing jitter with the root mean square (RMS) value of Δt_{RMS} as

$$\mathrm{SNR} = 10\log\left(\frac{A^2}{(A\cdot 2\pi f\cdot \Delta t)^2}\right) = -20\log(2\pi f\cdot \Delta t_{\mathrm{RMS}}) \tag{1.3}$$

Figure 1.2(b) shows a timing path of a typical digital logic, where t_{ck2q}, t_{delay}, and t_{set} represent the clock-to-Q delay of the flip-flop, the worst-case time delay of the combinational logic, and the setup time of the flip-flop, respectively. Assuming that the clock does not contain timing noise, we can prevent the setup-time violation when

$$t_{\mathrm{margin}} = T_{\mathrm{clk}} - t_{\mathrm{ck2q}} - t_{\mathrm{set}} - t_{\mathrm{delay}} > 0 \tag{1.4}$$

where t_{margin} and T_{clk} is the timing margin for the setup-time violation and period of the clock. If the timing noise Δt is introduced, (1.4) becomes

$$t_{\mathrm{margin}} = T_{\mathrm{clk}} - t_{\mathrm{ck2q}} - t_{\mathrm{set}} - t_{\mathrm{delay}} - \Delta t > 0 \tag{1.5}$$

which implies that the t_{margin} is degraded by Δt. Note that the timing noise also affects the hold-time violation in the same manner.

This book explains the various effects of jitter and its frequency-domain representation and phase noise, which are two different notational metrics that quantify the timing noise of a signal, and focuses on describing circuit techniques to achieve a low phase noise and jitter.

Chapter 2
Introduction to phase noise and jitter

2.1 Definition of jitter

Basically, jitter means timing variations of a periodic signal from ideal positions. More specifically, jitter refers to the short-term variations of a periodic signal's zero crossings from their ideal positions in time, whose frequency content exceeds 10 Hz [1]. On the other hand, the slow variations with a frequency content below 10 Hz are called wander and not considered as jitter [2].

There are several metrics for quantifying jitter. Sometimes definitions of jitter differ in the literature, so we must be careful of what definition the authors use. In this chapter, we start by defining time interval error (TIE), period jitter, and cycle-to-cycle jitter. Figure 2.1 shows the definitions of TIE, period jitter, and cycle-to-cycle jitter of a clock signal. TIE also has many different titles such as edge-to-edge jitter, time interval jitter, absolute jitter, phase jitter, or just jitter. TIE is defined as the absolute difference in the position of a clock's edge from the ideally exact position. Therefore, the ideal positions must be known or estimated to calculate TIE. On the other hand, the period jitter and cycle-to-cycle jitter do not need the ideal positions to be calculated. The period jitter, which is also called as cycle jitter, means the difference between any one measured clock period and the ideal clock period [3]. Although the period jitter definition refers to the ideal clock, its root of mean square (RMS) and peak-to-peak values are calculated statistically regardless of the ideal clock period. The period jitter (J_{period}) is expressed as

$$J_{\text{period}}(n) = P(n) - P_{\text{ideal}} \tag{2.1}$$

where P_n and P_{ideal} are the measured clock period at nth cycle and the ideal clock period, respectively. After having a variance operation to both sides of (2.1), it becomes

$$\text{var}(J_{\text{period}}) = \text{var}(P) \tag{2.2}$$

where P_{ideal} is a constant. As a result, the period jitter can be measured by observing variation on the second edge of the clock with the first edge being triggered, as

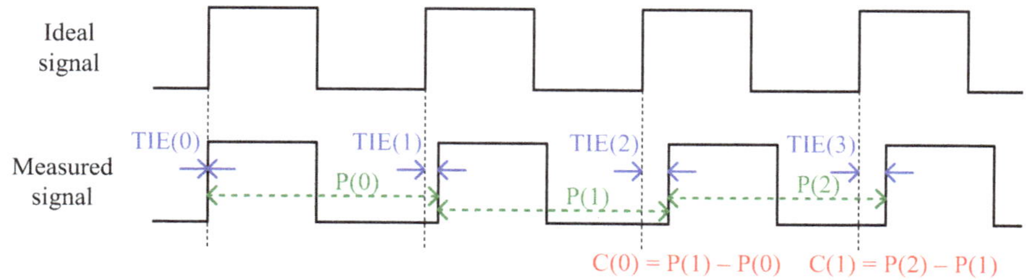

Figure 2.1 Time interval error (TIE), period jitter, and cycle-to-cycle jitter

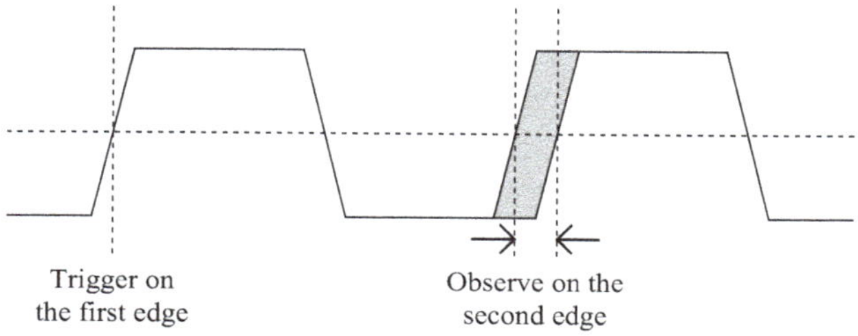

Figure 2.2 Measurement of the period jitter

shown in Figure 2.2. The cycle-to-cycle jitter (J_{c2c}) is defined as the difference between any two-adjacent clock period and can be expressed as

$$J_{c2c}(n) = \mathrm{P}(n+1) - P(n) \tag{2.3}$$

By substituting (2.1) into (2.3), (2.3) becomes

$$J_{c2c}(n) = J_{\mathrm{period}}(n+1) - J_{\mathrm{period}}(n) \tag{2.4}$$

which implies that the cycle-to-cycle jitter is a first-order difference of the period jitter. Therefore, the cycle-to-cycle jitter shows the instantaneous dynamics of the clock period. On the other hand, looking back at the TIE, it can also be expressed with P(n) and P_{ideal} as follows:

$$\mathrm{TIE}(n) = \sum P(n) - n \cdot P_{\mathrm{ideal}} \tag{2.5}$$

The first-order difference of both sides of (2.5) becomes

$$\mathrm{TIE}(n) - \mathrm{TIE}(n-1) = P(n) - P_{\mathrm{ideal}} \tag{2.6}$$

which is the same as the definition of period jitter in (2.1). That is, the TIE is the integration of the period jitter so that the TIE is appropriate for evaluating the cumulative effect of jitter over time.

2.2 Power spectral density

Simply put, the jitter and the phase noise refer to the same, but their difference comes from which domain they are expressed, time domain and frequency domain. Since the phase is not directly measurable whereas voltage or current wave can, it is required to have a conversion method that translates a time-domain voltage (or current) wave into the frequency domain. Therefore, a good understanding of power spectral density (PSD), which measures how the power of a signal or time series is distributed with frequency, is essential before studying the concept of phase noise. This chapter gives a brief review of PSD with a few simple examples to review the basics of PSD.

2.2.1 Pure sine wave

A pure sine wave is expressed as

$$y(t) = A_0 \sin(2\pi f_0 t + \phi_0) \tag{2.7}$$

where A_0, f_0, and ϕ_0 represent amplitude, frequency, and phase of the sine wave, respectively. By obtaining the Fourier transform from (2.7), we can get

$$Y(f) = \frac{1}{2}A_0 e^{i\phi_0}\delta(f - f_0) - \frac{1}{2}A_0 e^{-i\phi_0}\delta(f + f_0) \tag{2.8}$$

where $d(t)$ is a Dirac delta function. For the calculation of PSD, we will use Wiener–Khinchin theorem—the PSD of a wide-sense stationary random process is the Fourier transform of the corresponding autocorrelation function. Therefore, at first, we need to calculate autocorrelation function.

$$R(\tau) = E[y(t)y(t+\tau)] = \frac{1}{2}A_0{}^2\cos(2\pi f_0 \tau) \tag{2.9}$$

where $E(X)$ represents the expected value of random process X. Taking Fourier transform for both sides, (2.9) becomes

$$S_y(f) = \frac{1}{4}A_0{}^2\delta(f - f_0) + \frac{1}{4}A_0{}^2\delta(f + f_0) \tag{2.10}$$

Figure 2.3 illustrates the results of (2.8) and (2.10).

2.2.2 Sine wave with narrowband phase modulation

A sine wave with phase modulation is expressed as

$$y(t) = A_0 \sin(2\pi f_0 t + \phi_0 + \phi(t)) \tag{2.11}$$

where $\phi(t)$ is a time-varying phase modulation function. Assuming a narrowband phase modulation (PM), that is, the absolute amount of modulated phase is small

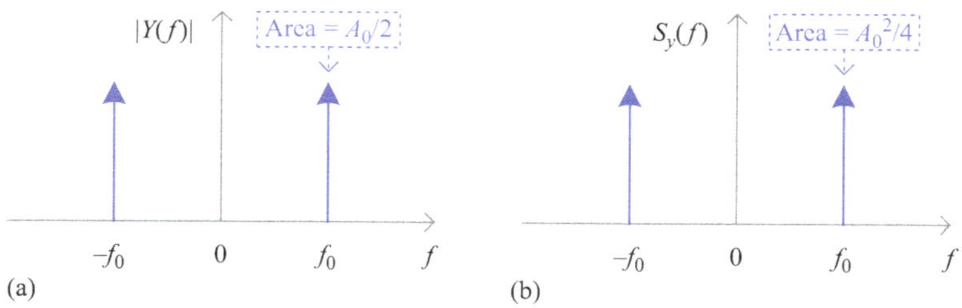

Figure 2.3 (a) Fourier transform of a pure sine wave and (b) PSD of a pure sine wave

enough (otherwise the modulation becomes frequency modulation (FM) and its analysis becomes more complex), (2.11) becomes

$$y(t) \cong A_0 \sin(2\pi f_0 t + \phi_0) + A_0\phi(t)\cos(2\pi f_0 t + \phi_0) \tag{2.12}$$

Because cos $\phi(t)$ and sin $\phi(t)$ are approximated to 1 and $\phi(t)$, respectively. The Fourier transform of $y(t)$ is

$$\begin{aligned} Y(f) = &\frac{1}{2}A_0 e^{i\phi_0}\delta(f-f_0) - \frac{1}{2}A_0 e^{-i\phi_0}\delta(f+f_0) + \frac{1}{2}A_0 e^{i\phi_0}\boldsymbol{\Phi}(f-f_0) \\ &- \frac{1}{2}A_0 e^{-i\phi_0}\boldsymbol{\Phi}(f+f_0) \end{aligned} \tag{2.13}$$

where $\boldsymbol{\Phi}(f)$ is the Fourier transform pair of $\phi(t)$. Note that the last two terms are added to (2.8). In a similar way, the autocorrelation of $y(t)$ is

$$\begin{aligned} R(\tau) &= \overline{y(t)y(t+\tau)} \\ &= \frac{1}{2}A_0^2\cos(2\pi f_0\tau) + \frac{1}{2}A_0^2\overline{\phi(t)\phi(t+\tau)}\cos(2\pi f_0\tau) \\ &= \frac{1}{2}A_0^2\cos(2\pi f_0\tau)\left(1 + R_\phi(\tau)\right) \end{aligned} \tag{2.14}$$

Fourier transform of (2.14) is

$$\begin{aligned} S_y(f) = &\frac{1}{4}A_0^2\delta(f-f_0) + \frac{1}{4}A_0^2\delta(f+f_0) + \frac{1}{4}A_0^2 S_\phi(f-f_0) \\ &+ \frac{1}{4}A_0^2 S_\phi(f+f_0) \end{aligned} \tag{2.15}$$

which implies that the narrowband component is modulated to the vicinity of the sine wave frequency. Note that the first two terms are the same as the PSD of the pure sine wave in (2.10), and each of the last two terms is the PSD of $\phi(t)$ scaled by $A_0^2/4$. Equation (2.15) is visualized in Figure 2.4(a).

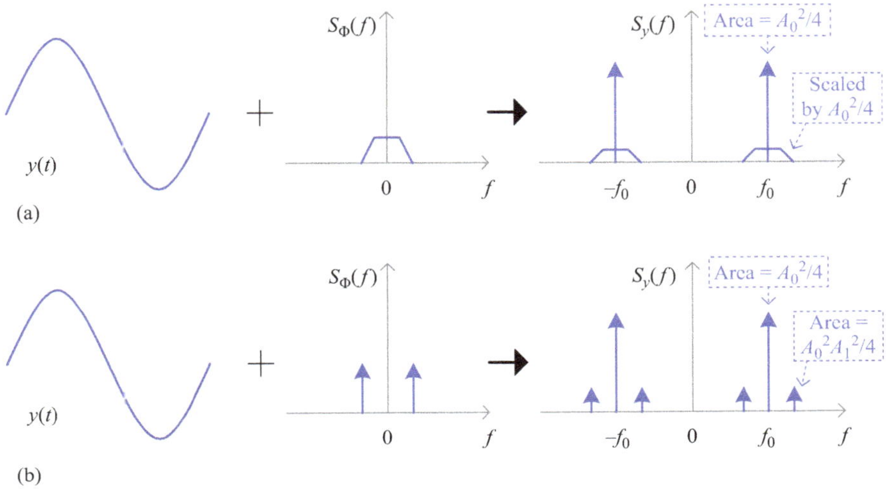

Figure 2.4 PSD of a sine wave (a) with narrowband phase modulation and (b) with sinusoidal phase modulation

For simplicity, if we assume a sinusoidal modulation for $\phi(t)$, the PSD of $\phi(t)$ becomes

$$S_\phi(f) = \frac{1}{4}A_1{}^2\delta(f - f_1) + \frac{1}{4}A_1{}^2\delta(f + f_1) \tag{2.16}$$

where A_1 and f_1 are amplitude and frequency of the sinusoidal modulation, respectively. The PSD of the sine wave with the sinusoidal modulation can be easily calculated from (2.15) and (2.16) to be

$$\begin{aligned} S_y(f) &= \frac{1}{4}A_0{}^2\delta(f - f_0) + \frac{1}{4}A_0{}^2\delta(f + f_0) \\ &\quad + \frac{1}{4}A_0{}^2\left(\frac{1}{4}A_1{}^2\delta(f - f_0 - f_1) + \frac{1}{4}A_1{}^2\delta(f - f_0 + f_1)\right. \\ &\quad \left. + \frac{1}{4}A_1{}^2\delta(f + f_0 - f_1) + \frac{1}{4}A_1{}^2\delta(f + f_0 + f_1)\right) \end{aligned} \tag{2.17}$$

Equation (2.17) is also visualized in Figure 2.4(b).

2.2.3 Sine wave with both jitter and noise

In this section, we will compare how white jitter and white noise contribute to the PSD of a sine wave. To be clearer, sine wave with white jitter means that the frequency profile of the phase modulation function $\phi(t)$ is "white", but sine wave with white noise means the white voltage is added on top of the pure sine wave. In the case of white jitter, $\phi(t)$ is white noise so that $R_\phi(\tau)$ equals to $a \cdot \delta(\tau)$, where a is the average noise level of the white noise. The autocorrelation of the signal is calculated from (2.14) to be

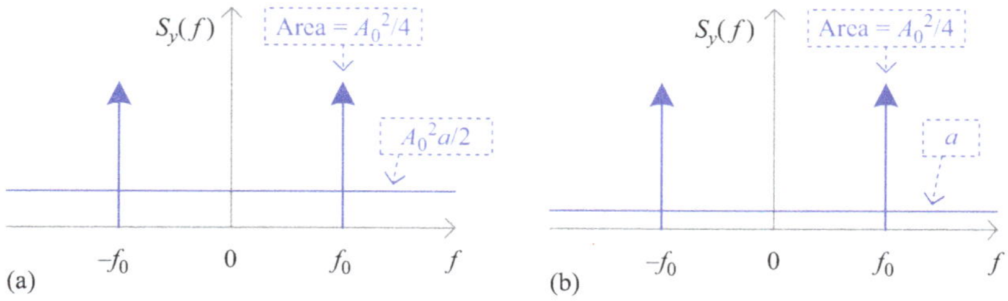

Figure 2.5 PSD of a sine wave (a) with white jitter and (b) with white noise

$$R(\tau) = \frac{1}{2}A_0^2\cos(2\pi f_0\tau)(1 + a\delta(\tau)) \tag{2.18}$$

Then the PSD becomes

$$S_y(f) = \frac{1}{4}A_0^2\delta(f - f_0) + \frac{1}{4}A_0^2\delta(f + f_0) + \frac{1}{2}A_0^2 a \tag{2.19}$$

We can find that signal-to-noise ratio (SNR) is a regardless of the signal amplitude A_0, because the noise floor from white jitter is proportional to A_0^2. On the other hand, in the case of white noise, the autocorrelation and the PSD are

$$R(\tau) = A_0^2\cos(2\pi f_0\tau) + a\delta(\tau) \tag{2.20}$$

$$S_y(f) = \frac{1}{4}A_0^2\delta(f - f_0) + \frac{1}{4}A_0^2\delta(f + f_0) + a \tag{2.21}$$

which means that the noise level is independent of the signal amplitude. As a result, SNR will improve as the signal amplitude increases. Figure 2.5(a) and (b) visualizes (2.19) and (2.21), respectively. Equation (2.19) gives an important nature of jitter, that is, increasing the signal power does not help to improve the SNR if the jitter is the dominant source of the SNR degradation. Note that we already observed this in the sample-and-hold circuit example given in (1.3). It highlights the significance of generating low-jitter clock in IC design.

2.3 Phase noise

Phase noise can be defined as frequency domain representation of rapid, short-term (i.e., >10 Hz), and random fluctuations in the phase of a periodic wave. As mentioned in Section 2.2, since it cannot be directly measured, PSD measurement is used to achieve the phase noise. Therefore, it is very important to understand the relation between PSD and phase noise of a periodic signal. Specifically, the phase

---

$$\mathrm{SNR} = 10\log\left(\frac{A^2}{(A\cdot 2\pi f\cdot\Delta t)^2}\right) = -20\log(2\pi f\cdot\Delta t_{\mathrm{RMS}}) \tag{1.3}$$

noise is characterized by the single-sideband noise spectral density, which is the ratio of the single sideband power at a frequency apart from the carrier frequency (f_0) by Δf over the total power under the power spectrum in logarithmic scale as follows [4].

$$\mathscr{L}(\Delta f) = 10\log\left(\frac{P_{\text{sideband}}(f_0 + \Delta f, 1\ \text{Hz})}{P_{\text{carrier}}}\right) \tag{2.22}$$

where $P_{\text{sideband}}(f_0 + \Delta f, 1\ \text{Hz})$ represents the single sideband power at $f_0 + \Delta f$ in a measurement bandwidth of 1 Hz. Note that the measurement bandwidth should be introduced because the limit of P_{sideband} becomes zero as Δf approaches zero. $\mathscr{L}(\Delta f)$ is pronounced "script-ell of delta f." When expressed in decibels, the unit of $\mathscr{L}(\Delta f)$ is decibels below the carrier in a 1 Hz bandwidth (dBc/Hz) [5].

Figure 2.6 illustrates how to convert PSD to phase noise, with an example of a sine wave with a narrowband sinusoidal modulation. As described in Section 2.2, when a sinusoidal component ($\phi(t) = A_1 \sin(2\pi f_1 t + \phi_1)$) is modulated, four spurious tones occurs the PSD of the sine wave ($y(t) = A_0 \sin(2\pi f_0 t + \phi_0 + \phi(t))$) at the frequencies of $-f_0 \pm f_1$ and $f_0 \pm f_1$, whose magnitudes are $A_0^2 A_1^2/16$. From the PSD of the signal, we can easily convert the double-sideband (DSB) PSD to the single-sideband (SSB) form, where power of each frequency component is doubled. In practice, the SSB PSD result $S_{y,\text{SSB}}(t)$ can be measured when we connect the periodic signal to a spectrum analyzer. After that, we can achieve the SSB noise spectral density, $S_{y,\text{SSB}}(\Delta f)$, which shows normalized power and frequency of the

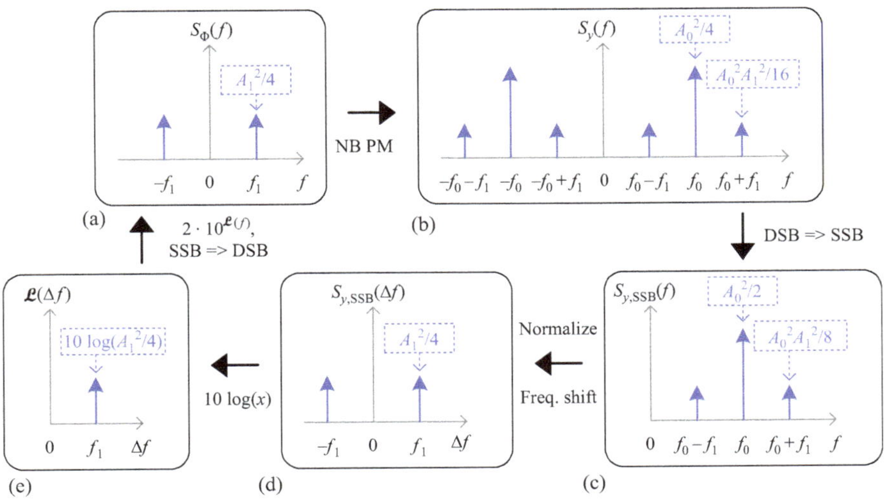

Figure 2.6 Relation between PSD and phase noise of a sine wave with a sinusoidal modulation: (a) PSD of $\phi(t)$, (b) PSD of $y(t)$, (c) SSB conversion of (b), (d) SSB PSD down-converted and normalized to f_0, and (e) the phase noise

spurious tones relative to the power and frequency of the carrier signal. The phase noise $\mathscr{L}(\Delta f)$ is defined as the SSB noise spectral density with logarithmic scale on the positive Δf side, which is shown in Figure 2.6(e). Note that a simple DSB-to-SSB conversion and logarithmic expression (Figure 2.6(a) → Figure 2.6(e)) can achieve the phase noise from $S(f)$ as

$$L(\Delta f) = 10\log\left(\frac{1}{2}S_\phi(\Delta f)\right) \tag{2.23}$$

In practice, however, the PSD of the signal is the only one measurable so that we must follow (b) → (c) → (d) → (e) to get $\mathscr{L}(\Delta f)$. PSD to phase noise conversion flow for a general phase noise profile is shown in Figure 2.7.

2.4 Relation of phase noise and jitter

In this section, the relation of the phase noise and jitter will be studied. There are two main types of jitter: synchronous jitter and accumulating jitter. The synchronous jitter can be simply defined as undesired variation in the delay between the input and the produced output. Therefore, it does not have any memory or accumulation effect. In terms of the PSD, the synchronous jitter exhibits a white property. On the other hand, the accumulating jitter comes from an accumulation

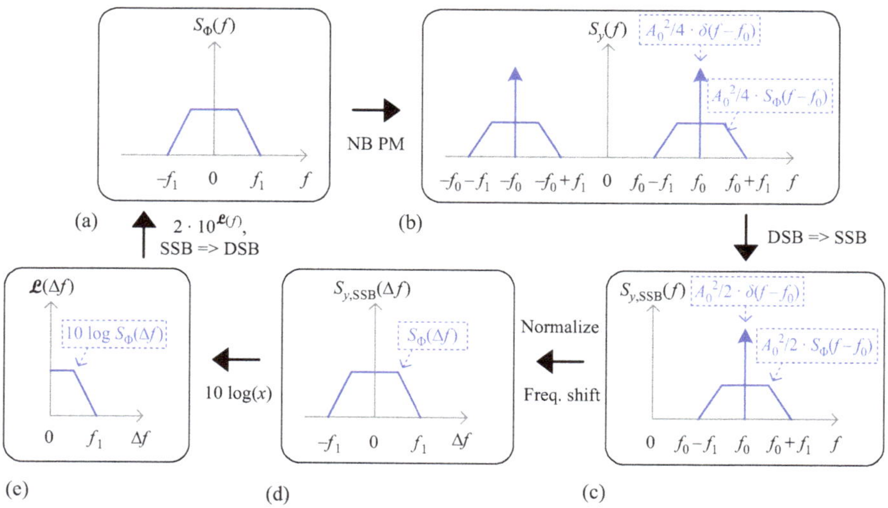

Figure 2.7 Relation between PSD and phase noise of a sine wave with a narrowband modulation: (a) PSD of $\phi(t)$, (b) PSD of $y(t)$, (c) SSB conversion of (b), (d) SSB PSD down-converted and normalized to f_0, and (e) the phase noise

of all variations in the delay between an output transition and the subsequent output transition. That is, a timing variation in one cycle is added to sum of all the previous cycles and affects the subsequent cycles. In CMOS circuits, a signal buffer which adds only a delay in the signal path introduces the synchronous jitter, but an oscillator where a disturbance in the current cycle circulates to subsequent cycles mainly introduces the accumulation jitter. An example illustration of the synchronous jitter and the accumulating jitter in time domain is shown in Figure 2.8. The PSD of accumulation jitter is proportional to $1/f^2$ due to the integrating nature. The detailed derivation and comparison of the jitters by the buffer and the oscillator will be studied in Chapter 4. This chapter focuses on how the synchronous jitter and the accumulation jitter are obtained from (converted to) the phase noise.

From the definition of the period jitter, the variance of the period jitter can be expressed as

$$\sigma_{\text{per}}^2 = \frac{E\left[(\phi(t+T) - \phi(t))^2\right]}{(2\pi f_0)^2}$$
$$= \frac{E[\phi^2(t+T)] - 2E[\phi(t)\phi(t+T)] + E[\phi^2(t)]}{(2\pi f_0)^2} \tag{2.24}$$

where $\frac{1}{2\pi f_0}$ is multiplied to convert phase to time. From (2.9), (2.24) can be rewritten as

$$\sigma_{\text{per}}^2 = \frac{2\left(R_\phi(0) - R_\phi(T)\right)}{(2\pi f_0)^2} \tag{2.25}$$

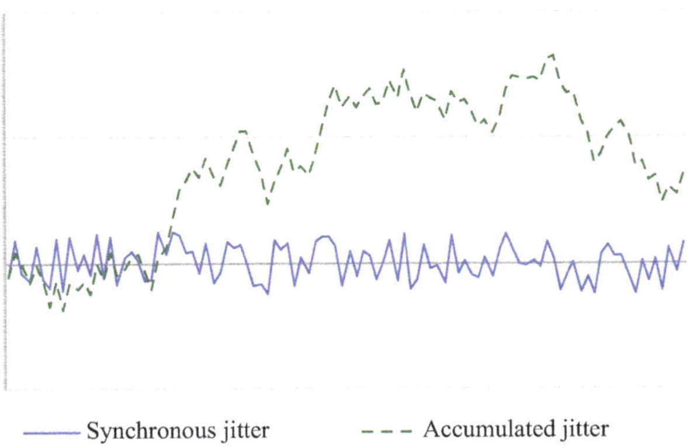

Figure 2.8. An example of synchronous jitter and accumulated jitter in time domain

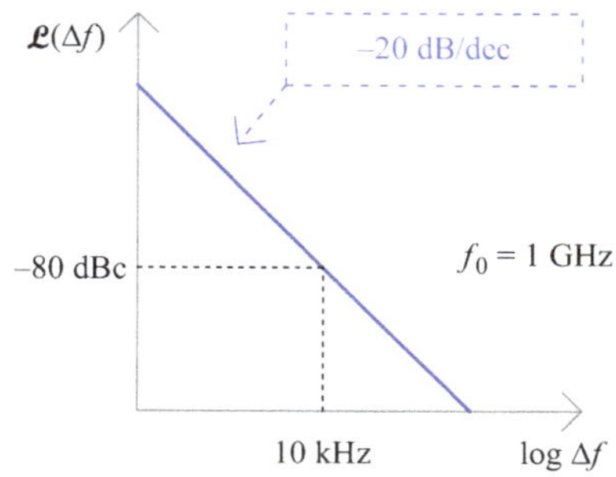

Figure 2.9 Oscillator phase noise example

Since the PSD and the autocorrelation of a signal are the Fourier transform pair, (2.25) leads to the following relation between the period jitter and the PSD:

$$\sigma_{\text{per}}^2 = \frac{2\left(\int_{-\infty}^{\infty} S_\phi(f)df - \int_{-\infty}^{\infty} S_\phi(f)e^{j2\pi fT}df\right)}{(2\pi f_0)^2}$$
$$= \frac{2\left(\int_{-\infty}^{\infty} S_\phi(f)(1-\cos(2\pi fT))df\right)}{(2\pi f_0)^2} = \frac{4\left(\int_{-\infty}^{\infty} S_\phi(f)\sin^2(\pi fT)df\right)}{(2\pi f_0)^2} \tag{2.26}$$

When we consider the accumulation jitter, the PSD can be written as

$$S_\phi(f) = c\frac{f_0^2}{f^2} \tag{2.27}$$

Combined with (2.27), (2.26) becomes

$$\sigma_{\text{per}}^2 = \frac{4\left(\int_{-\infty}^{\infty} cf_0^2 \cdot \frac{\sin^2(\pi fT)}{f^2}df\right)}{(2\pi f_0)^2} = \frac{4\cdot cf_0^2\cdot\pi^2 T}{(2\pi f_0)^2} = cT \tag{2.28}$$

From (2.28), the period jitter of an oscillator can be easily obtained from the measure PSD or phase noise. For example, in the case of an oscillator example whose phase noise is given as Figure 2.9, the constant of the PSD is obtained by solving:

$$S_\phi\,(10\text{ kHz}) = c\frac{(1\text{ GHz})^2}{(10\text{ kHz})^2} = 10^{-8} \tag{2.29}$$

which leads to $c = 10^{-18}$. As a result, the RMS period jitter of the oscillator becomes

$$\sigma_{\text{per}} = \sqrt{cT} = 31.6\text{ fs} \tag{2.30}$$

On the other hand, the accumulation jitter is calculated from the phase noise more intuitively. The RMS value of $\phi(t)$ is obtained from the integration of $S_\varphi(f)$ as

$$\phi_{\rm rms} = \sqrt{\int_{f_{\min}}^{f_{\max}} S_\phi(f)df} \tag{2.31}$$

where $f_{\min}$ and $f_{\max}$ denote the frequency range of interest. Substituting (2.23), (2.31) is rewritten as

$$\phi_{\rm rms} = \sqrt{2\cdot\int_{f_{\min}}^{f_{\max}} 10^{\mathscr{L}(\Delta f)}df} \tag{2.32}$$

The RMS jitter can be obtained from (2.32) by transforming the phase domain to the time domain:

$$\sigma_{\rm TIE} = \frac{1}{2\pi f_0}\sqrt{2\cdot\int_{f_{\min}}^{f_{\max}} 10^{\mathscr{L}(\Delta f)}df} \tag{2.33}$$

References

[1] EXFO. *FTB-8080 Sync Analyzer: Resolving Synchronization Problems in Telecom Networks*. Application Note 119, 2012.

[2] Tektronix. *Understanding and Characterizing Timing Jitter*. 2002.

[3] Maxim Integrated. *Clock (CLK) Jitter and Phase Noise Conversion*. Application Note 3359, 2004.

[4] Hajimiri A, Lee TH. A general theory of phase noise in electrical oscillators. *IEEE Journal of Solid-State Circuits*. 1998;33(2):179–194.

[5] Vig JR. IEEE standard definitions of physical quantities for fundamental frequency and time metrology-random instabilities. *IEEE Standard*. 1999; 1139:1999.

Chapter 3
CMOS oscillators

In definition, an electronic oscillator is an electronic system consisting of active and passive circuit elements to produce a periodic signal at the output without the application of an external input signal [1]. Because an oscillator itself generates a periodic signal, specifically a clock signal, the spectral purity of the signal highly depends on the quality of the oscillator. Even a standalone oscillator can serve as a clocking circuit depending on the system requirement. Therefore, a CMOS oscillator is the most important building block of a CMOS clocking circuit that should be studied in depth. This chapter will introduce fundamentals of two popular CMOS oscillator topologies: LC oscillator and ring oscillator.

3.1 LC oscillator

Among the three circuit elements (resistor, capacitor, and inductor), integrated inductors in CMOS technology has the shortest history than the others, as it is a bit harder to be monolithically integrated into a chip because of its huge size and many parasitic effects [2–6]. For the last 30 years, there have been many pioneer works that try to enable high-quality integrated inductors [2–11]. Thanks to those efforts, nowadays, it is easy to find on-chip inductors in a variety of applications.

Ideally a simple LC tank shown in Figure 3.1 can be an oscillator. The impedance of parallel LC elements equals:

$$Z_{\mathrm{LC}} = \frac{j\omega L \cdot \frac{1}{j\omega C}}{j\omega L + \frac{1}{j\omega C}} = \frac{j\omega L}{1 - \omega^2 \mathrm{LC}} \tag{3.1}$$

where ω is the angular frequency. We can find that the impedance becomes infinite when $\omega = \frac{1}{\sqrt{LC}}$, which implies that the parallel LC circuit resonates at that frequency. Because there is no lossy element, the initial energy stored in the inductor and capacitor will not be lost forever so that the oscillation will sustain. In reality, however, the inductor and capacitor cannot be ideal. There should be a resistive (lossy) element, which is generally more severe for the inductor when integrated into a chip. The parasitic resistance to the inductor is typically a series resistance, but it can be approximately converted to the parallel resistance as shown in Figure 3.1(b), with a transformation of $R = (1 + Q^2)R_S$ [12,13], where R_S is the series resistance of the inductor and Q is the quality factor of the inductor. The

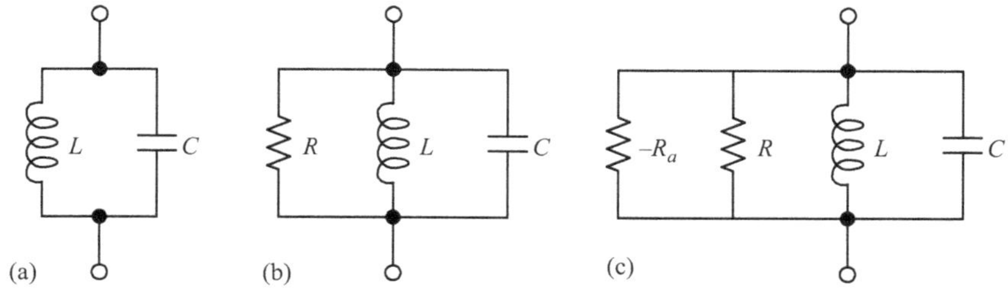

Figure 3.1 (a) Lossless LC tank, (b) lossy LC tank, and (c) lossy LC tank with a negative resistance

detailed transformation of the series resistance to the parallel resistance is given in Section 3.3. Applying Kirchhoff's current law (KCL) leads to

$$I_R + I_L + I_C = \left(\frac{V(t)}{R}\right) + \left(V(0) + \int \frac{V(t)}{L}dt\right) + \left(\frac{d}{dt}CV(t)\right) = 0 \tag{3.2}$$

where $V(0)$ is the initial voltage of the RLC circuit. By taking the time derivative and dividing by C, (3.2) becomes

$$\frac{d^2}{dt^2}V(t) + \frac{1}{\mathrm{RC}}\frac{d}{dt}V(t) + \frac{1}{\mathrm{LC}}V(t) = 0 \tag{3.3}$$

When we define $\omega_0 = \frac{1}{\sqrt{LC}}$ and $\mathrm{Q} = R\sqrt{\frac{C}{L}}$ (3.3) becomes

$$\frac{d^2}{dt^2}V(t) + \frac{\omega_0}{Q}\frac{d}{dt}V(t) + \omega_0^2 V(t) = 0. \tag{3.4}$$

The general solution of (3.4) is

$$V(t) = V_1 \mathrm{e}^{-\omega_0\left(\frac{1}{2Q}+\sqrt{\left(\frac{1}{2Q}\right)^2-1}\right)t} + V_2 \mathrm{e}^{-\omega_0\left(\frac{1}{2Q}-\sqrt{\left(\frac{1}{2Q}\right)^2-1}\right)t} \tag{3.5}$$

If Q > 1/2, the root terms become imaginary, and therefore the right side of (3.5) becomes

$$V(t) = V_1 \mathrm{e}^{-\frac{\omega_0}{2Q}t}\cos\left(\omega_0\sqrt{1-\left(\frac{1}{2Q}\right)^2}\,t\right) + V_2 \mathrm{e}^{-\frac{\omega_0}{2Q}t}\sin\left(\omega_0\sqrt{1-\left(\frac{1}{2Q}\right)^2}\,t\right) \tag{3.6}$$

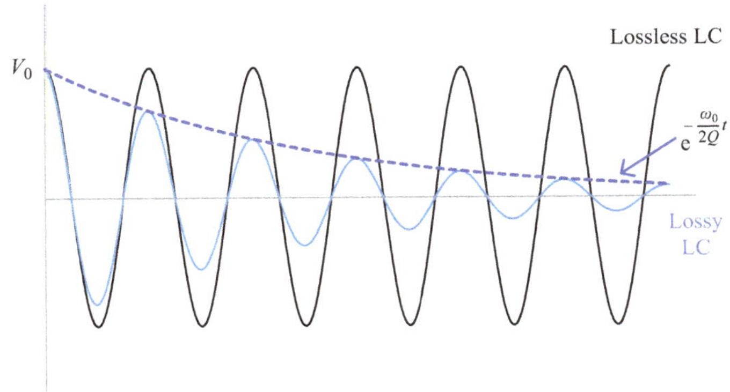

Figure 3.2 Lossy LC resonance waveforms

which can be simplified as

$$V(t) = V_0 e^{-\frac{\omega_0}{2Q}t} \sin\left(\omega_0 \sqrt{1 - \left(\frac{1}{2Q}\right)^2}\, t + \arctan\left(\frac{V_2}{V_1}\right)\right) \tag{3.7}$$

Equation (3.7) has a sinusoidal term but the amplitude decreases with time constant of $2Q/\omega_0$, which means that the oscillation is not sustainable. Figure 3.2 compares (3.7) over the output waveform of the lossless LC tank.

From (3.7), we can infer that the resistance R should be canceled out by introducing a negative resistance $-R_a$ as shown in Figure 3.1(c), to implement a practical LC oscillator. Simply, if $R_a > R$, the total resistance is positive, then the LC tank is still lossy. If $R_a = R$, the total resistance becomes zero and the exponential term equals unity, so the LC tank will sustain a constant voltage swing. If $R_a < R$, the total resistance and Q become negative. When we substitute negative Q into (3.7), the amplitude of sine wave expands exponentially, and eventually reaches infinite, as shown in Figure 3.3.

In practice, the negative resistance can be implemented with a transconductance (g_m) element, as shown in Figure 3.4. The resistance of the g_m element equals $-1/g_m$, so a well-controlled g_m can compensate R. The amplitude of oscillation exponentially increases, and its energy will become infinite if $1/g_m < R$, which is definitely not feasible in the real world. In fact, the g_m of a practical device saturates as the applied voltage (V) increases. That is, although the g_m is enough to compensate the loss due to R for a lower voltage, it eventually becomes smaller as the voltage increases. Figure 3.5 shows how the voltage swing of this LC tank is set, where V–I curves of the g_m element and the resistor are shown. For a low voltage region, the g_m is higher than $1/R$, and the area between the two curves represents the excess power (power supplied from $-1/g_m$ – power dissipated by R) because the integration of V–I curve is power. In the same manner, the area

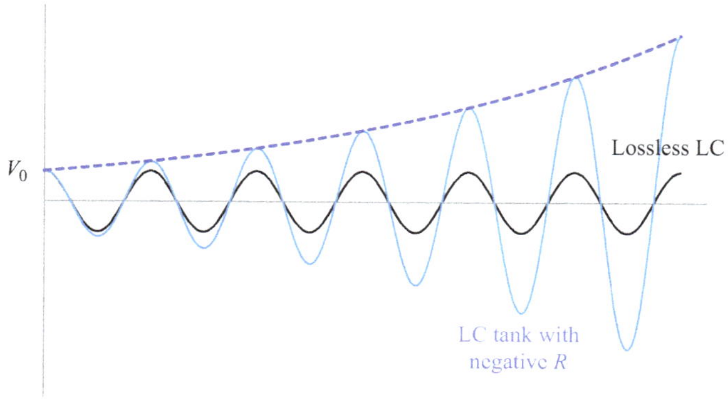

Figure 3.3 LC resonance waveform with negative total resistance

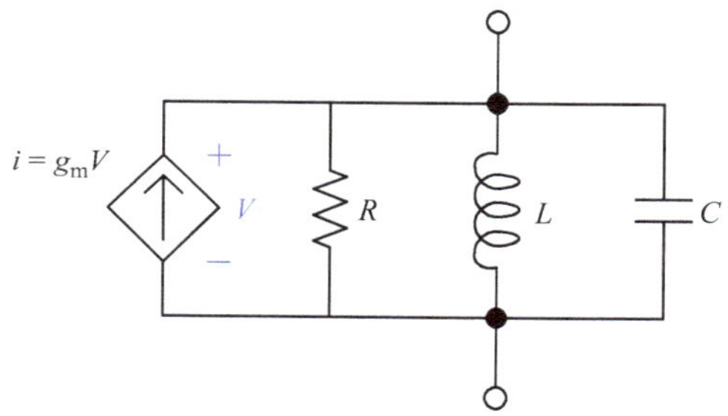

Figure 3.4 Lossy LC tank with a g_m element

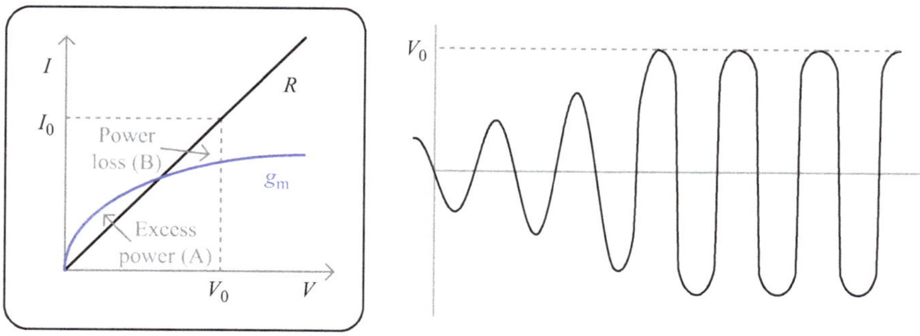

Figure 3.5 LC oscillator with nonlinear g_m

between the two curves at a higher voltage region is the power loss. It is obvious that the equilibrium is met when the excess power equals the power loss; therefore, the voltage swing V_0 is set where makes area A and area B be the same. In other words, the average g_m across the range below V_0 equals $1/R$.

Item	LC	Ring
Tuning range	~20%	~100%
Phase noise	Good	Poor
Size	Large	Small
Freq.	High	Low
Multiphase	Difficult	Easy

Figure 3.6 Comparison of LC and ring oscillators

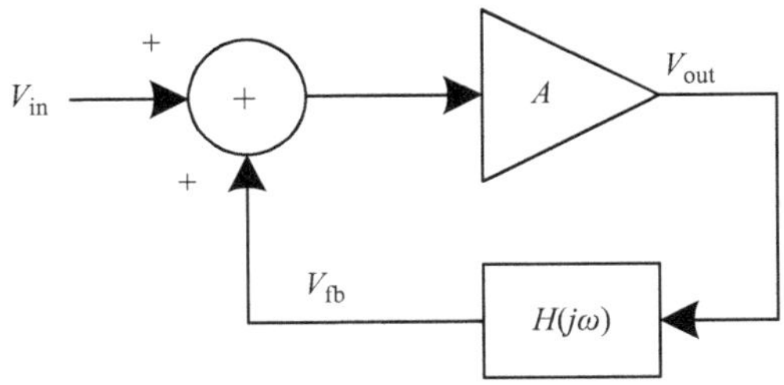

Figure 3.7 Simplified block diagram of feedback system

3.2 Ring oscillator

Ring oscillator is widely used in many applications because it has many distinctive advantages over the LC oscillator: it provides compact size, wide tuning range, and ease of multiphase generation. On the other side, phase noise and frequency of ring oscillator are much worse than those of LC oscillators, which limit the use of ring oscillator to low-frequency and less-sophisticated applications. A rough comparison of ring oscillator and LC oscillator is provided in Figure 3.6.

To understand ring oscillator, Barkhausen stability criterion should be mentioned. The Barkhausen criterion determines the condition when a feedback system can oscillate. For a simple feedback system shown in Figure 3.7, the closed-loop gain V_{out}/V_{in} can be calculated by solving:

$$A(V_{in} + V_{fb}) = V_{out} \tag{3.8}$$

Substituting $V_{fb} = H(j\omega)V_{out}$ into (3.8) leads to

$$\frac{V_{out}}{V_{in}} = \frac{A}{1 - AH(j\omega)} \tag{3.9}$$

We can observe from (3.9) that if A and $H(j\omega_0)$ meet the condition of the denominator equals zero. Note that they are the open-loop gain and the open-loop phase shift of the feedback loop, respectively. The closed-loop gain becomes infinite and the system does not converge at the frequency of ω_0, that is it will oscillate. Equation (3.10) is necessary but not sufficient conditions for sustainable oscillations (because it can stay at equilibria point if there is no stimulus) and is referred to as the Barkhausen stability criterion [14]. The criterion is visualized in Figures 3.8 and 3.9. If the phase shift by the loop meets the criteria, V_{in} and V_{fb} are fully in phase so that their amplitudes are added, as shown in Figure 3.8(a), and therefore the amplitude grows persistently. On the other hand, the signal amplitude becomes smaller when the phase shift equals π since the signs of V_{in} and V_{fb} are opposite. In fact, as defined in at the introduction of this chapter, an oscillator should oscillate without application of external signal. Figure 3.9 explains how oscillation grows in a feedback system satisfying (3.10), in the absence of an external signal. When the oscillator is turned on, there are many possible sources such as electronic noise of the circuit elements or power supply transient, which is able to insert an initial excitation into the oscillator [14–16]. If this momentary excitation contains the frequency component at ω_0, it is amplified by A during the first cycle after the excitation. Note that frequency component of such excitation usually distributes over a wide range, so it contains ω_0 component at a high probability. The amplified signal turns back to the starting point through the feedback path. From that, even though the initial excitation has gone, the returned signal becomes another "stronger" excitation, and it will propagate and grow while running the loop like a snowball. Because the signal is amplified by $|A \cdot H|$ for each lap, it is found that $|A \cdot H| \geq 1$ is a necessary condition for sustaining oscillation once again.

$$|AH(j\omega_0)| = 1, \ \angle AH(j\omega_0) = 2\pi \tag{3.10}$$

From now on, let us study how a CMOS ring oscillator works, based on the Barkhausen criterion. Basically, a chain of gain stages whose output is fed back to the input is a possible ring oscillator. For example, we can use a common-source (CS) gain stage which is shown in Figure 3.10(a). The transfer function of the CS amplifier is given as

$$A(j\omega) = -\frac{g_m R_L}{1 + j\omega R_L C_L} \tag{3.11}$$

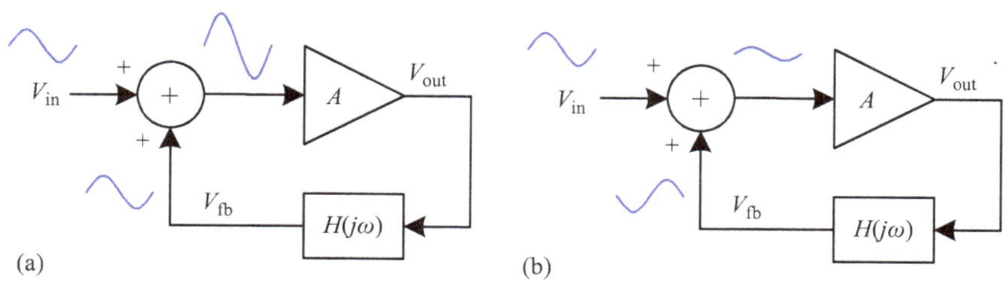

Figure 3.8 Feedback loop (a) with phase shift of 2π and (b) with phase shift of π

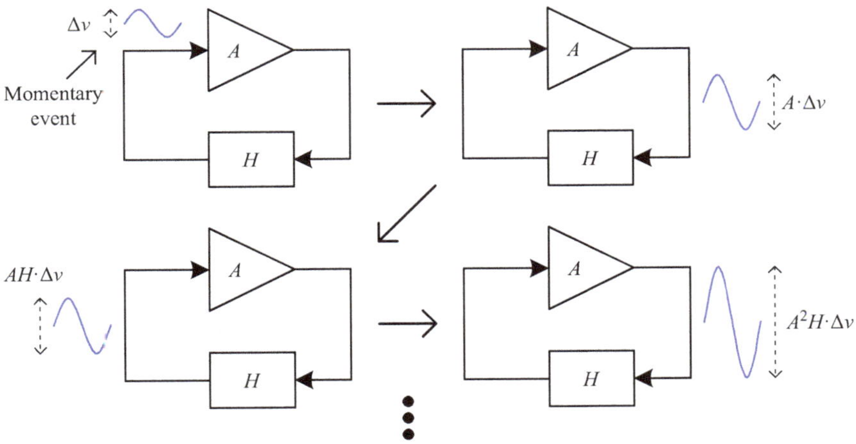

Figure 3.9 Buildup of oscillation with time when an oscillator meets the Barkhausen criterion

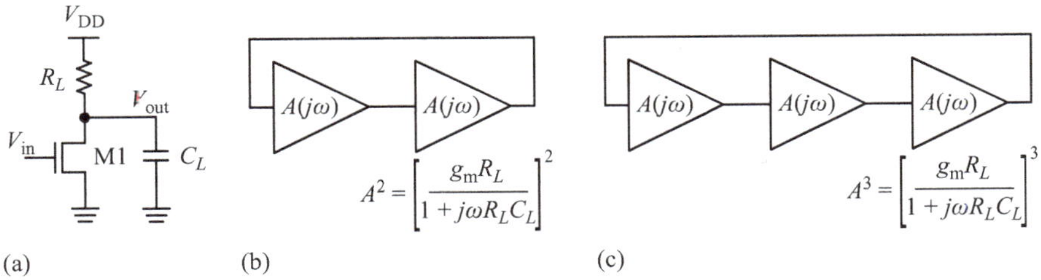

Figure 3.10 (a) Common-source amplifier, (b) open-loop gain of two-stage CS stages, and (c) open-loop gain of three-stage CS stages

where g_m, R_L, and C_L are the transconductance of N-type metal-oxide-semiconductor (NMOS) transistor M1, load resistance, and load capacitance. Equation (3.11) has only one pole, and therefore the phase shift due to the pole is less than $\pi/2$ regardless of the gain. Noting that the polarity of DC gain is negative, the overall phase shift of the CS stage is always within the range of π to $3\pi/2$, which means a single CS stage cannot meet the Barkhausen criterion. When we make a ring with two-stage CS stages as shown in Figure 3.10(b), the open-loop gain equals:

$$(A(j\omega))^2 = \left(\frac{g_m R_L}{1 + j\omega R_L C_L}\right)^2 \tag{3.12}$$

Because of the even number of stages, the DC gain is positive. That means the additional phase shift of π which comes from DC polarity is not available in this case so that the maximum phase shift of the two-stage configuration is less than π. It forms a positive feedback at a low frequency, so it is more like a latch. It is the

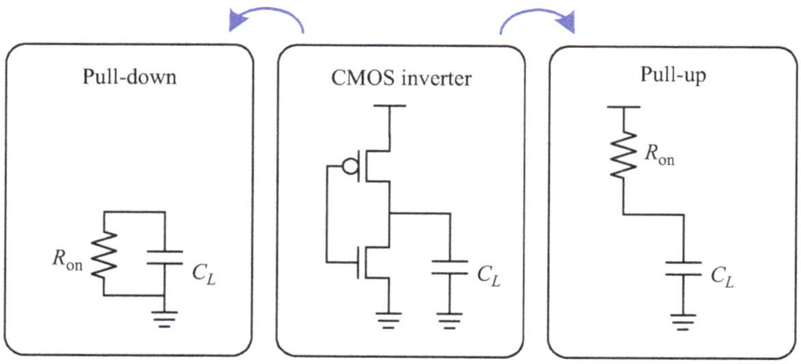

Figure 3.11 CMOS inverter RC model

main reason that a ring oscillator typically has odd number of stages. Moreover, even if we have the polarity shift by inserting an ideal inverting stage, the overall phase shift still cannot be 2π, unless the DC gain $g_m R_L$ is infinite. On the other hand, for the three-stage configuration shown in Figure 3.10(c), the required phase shift per stage is $\pi/3$, which is available in a single-pole transfer function. The frequency satisfying the phase shift of $\pi/3$ is calculated from the denominator of (3.11) as

$$\omega_0 = \frac{\tan\left(\frac{\pi}{3}\right)}{R_L C_L} = \frac{\sqrt{3}}{R_L C_L} \tag{3.13}$$

Substituting (3.13) into (3.11), the magnitude of the denominator becomes 2. Therefore, to meet the Barkhausen criteria for gain, $g_m R_L$ should be larger than or equal to 2. If the loop gain is larger than unity, the amplitude will grow exponentially and eventually saturates in a real circuit, where the average gain becomes unity [12]. In other words, the three-stage configuration can oscillate as long as the DC gain of CS stage is no less than 2, with the frequency of $\frac{\sqrt{3}}{R_L C_L}$.

CMOS inverters can also be used as a gain stage, which is the most common way of implementing a ring oscillator in recent CMOS technologies. As shown in Figure 3.11, a CMOS inverter is represented by a simple RC combination so that it can be regarded as a single-pole stage. Therefore, similar to the CS gain stage, at least three stages are required to implement a ring oscillator. Figure 3.12 shows simplified voltage and current waveforms of a three-stage ring oscillator based on CMOS inverters. Let us start from the first rising edge of the output of the inverter A. When it reaches $V_{DD}/2$ (crossover voltage of CMOS inverter), the driving strength of NMOS transistor in the inverter B surpasses that of P-type metal-oxide-semiconductor (PMOS), and then the output of the B starts discharging from V_{DD}. Note that we assume a first-order linear transition for simplicity, although the actual transition is a complex higher-order function of the input and output voltages. We can define the time difference between the zero-crossing time of A output to that of B output as the inverter delay (τ). Then, each inverter is

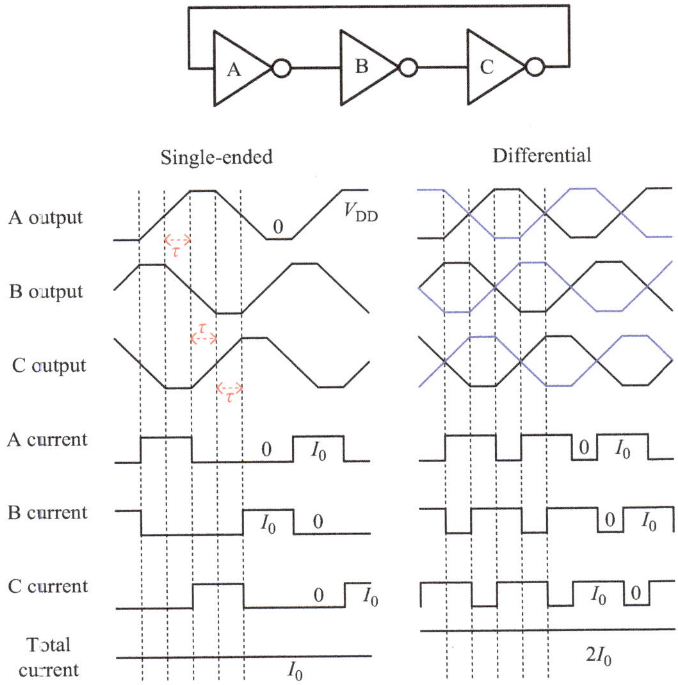

Figure 3.12 Voltage and current waveforms of inverter-based ring oscillator

triggered every 3τ, so that its period equals 6τ. We can easily notice that the oscillation period of an N-stage ring oscillator is $2N\tau$. When we look at the current consumption of a single CMOS inverter, it does not flow a static current. The inverter draws the current only when the output is transitioning, that is, the PMOS draws the current from V_{DD} to charge the output when the input is low but the NMOS discharges it to ground while the input is switched to high. From the viewpoint of V_{DD} port, the inverter draws current only at the output low-to-high transition. Therefore, assuming the transition time of 2τ, the current consumption of each inverter in the ring oscillator sequentially draws current I_0 from V_{DD} as shown in Figure 3.12, and therefore the total current is a constant, which is the same as the dynamic current of each inverter (I_0). On the other hand, if a differential configuration is used, each inverter draws current at every transition regardless of the polarity. Therefore, the total current becomes $2I_0$ but it remains constant when we neglect the additional current consumption (i.e., cross-coupled latch) for differential signaling. The differential configuration is usually implemented by coupling two inverters using a cross-coupled inverter pair, as shown in Figure 3.13. Owing to the nature of differential signaling, the differential ring oscillators provide a better sensitivity to common-mode noises such as power supply noise [17,18]. On the other hand, with the differential configuration, the ring

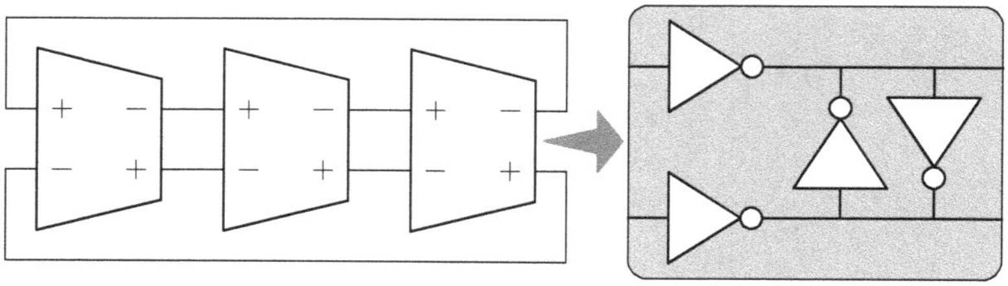

Figure 3.13 Three-stage CMOS pseudo-differential ring oscillator

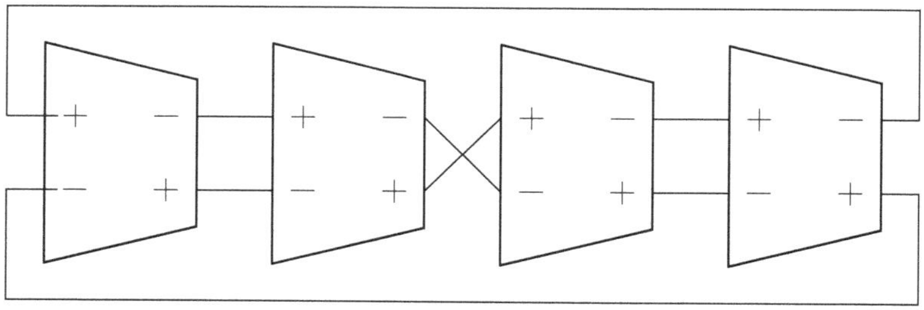

Figure 3.14 Four-stage differential ring oscillator

oscillator with an even number of stages can easily be enabled. That is, a differential configuration inherently provides the ideal inverting stage which we mentioned above. By twisting one of the connections between the delay elements, the polarity of feedback is flipped so that the latch-up at DC is prevented. Figure 3.14 shows a four-stage differential ring oscillator. We can find that the oscillation period is 8τ, which means that the phase shift per delay element corresponds to $\pi/4$ with respect to the oscillation period. Multiphase clock whose phases are evenly apart by $2\pi/2^N$, which is frequently required in many applications, is achieved from the ring oscillator that has even number of stages.

We have learnt that the frequency of a ring oscillator is determined by the delay of each delay element and the number of delay stages used therein. If identical inverters are used, it seems that each inverter has fanout-of-1 (FO1) loading. It can mislead that the speed of a ring oscillator is independent of the inverter sizing. In practice, parasitic capacitance introduced by the routing wire between adjacent cells sets the speed of a ring oscillator. As shown in Figure 3.15, if we have wire capacitance of C_{wire} while the input capacitance of each inverter is C_{in}, the fanout becomes

$$\text{FO} = \frac{C_{\text{in}} + C_{\text{wire}}}{C_{\text{in}}} \tag{3.14}$$

For delay calculation, note that drain capacitance γC_{in} should be considered as well, where γ (typically $0.5 < \gamma < 1$, $\gamma \approx 1$ in modern CMOS process) is the ratio of the drain capacitance over the gate capacitance of MOS transistor, which is

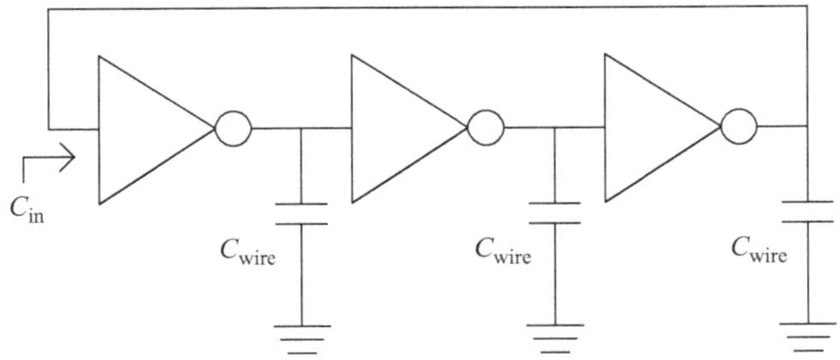

Figure 3.15 Ring oscillator with wire capacitance

determined by the process. From the fact that the C_{wire} has a dependency on C_{in} in general, C_{wire} can be expressed as

$$C_{wire} = C_{wire0} + \alpha C_{in} \tag{3.15}$$

where $0 < \alpha < 1$. Substituting (3.15) into (3.14), we can get:

$$FO = \frac{C_{in}(1+\alpha) + C_{wire0}}{C_{in}} \tag{3.16}$$

If though extremely large inverters are used, that is, $C_{in} \gg C_{wire0}$, (3.16) yields the minimum achievable fanout of $1 + \alpha$. It can be observed that the frequency of ring oscillator strongly depends on layout strategy, and we will get a wrong number if the wire parasitic is not correctly considered.

3.3 Appendix: Translation of series R-L to parallel R-L

In Section 3.1, we assumed that the resistance is in parallel with the inductor. However, in practice, the parasitic resistance of an integrated inductor originates to the metal resistance of the inductor, which is placed in series with the inductance as shown in the left side of Figure 3.16. However, because of the parallel capacitance, the series R-L network makes the RLC analysis a bit complex, compared to the parallel RLC tank analysis that we discussed in Section 3.1. As a remedy to that, we can use an equivalent parallel R-L network instead of the series R-L. In this appendix, we derive the translation of the parallel R-L network to the equivalent series R-L network.

Basically, to be an equivalence, the impedance of the parallel R-L network should equal to that of the series R-L. Their impedances are expressed as

$$Z_p = \frac{j\omega R_p L_p}{R_p + j\omega L_p} = \frac{\omega^2 R_p L_p^2 + j\omega R_p^2 L_p}{R_p^2 + \omega^2 L_p^2} \tag{3.17}$$

$$Z_s = R_s + j\omega L_s \tag{3.18}$$

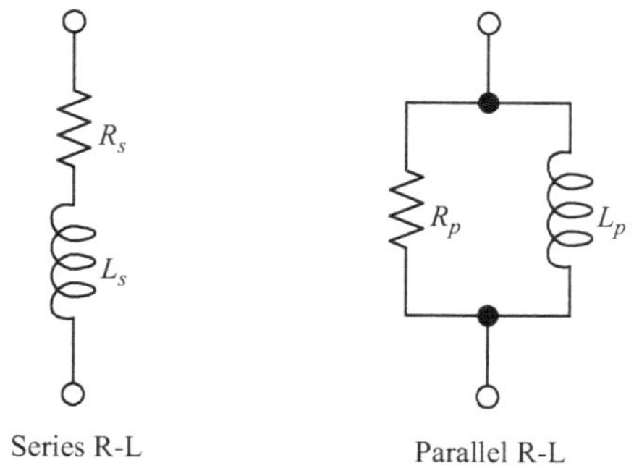

Figure 3.16 Series R-L and parallel R-L

where Z_p, Z_s, R_p, R_s, L_p, and L_s represent the impedance, resistance, and inductance of parallel/series networks, respectively. To be equivalent, the real and imaginary parts of (3.17) and (3.18) should equal each other, which are written as

$$Z_{\text{real}} = \frac{\omega^2 R_p L_p^2}{R_p^2 + \omega^2 L_p^2} = R_s \tag{3.19}$$

$$Z_{\text{imag}} = \frac{j\omega R_p^2 L_p}{R_p^2 + \omega^2 L_p^2} = j\omega L_s \tag{3.20}$$

At the same time, there is another requirement for the equivalency, that is, the quality factor. That is, the ratios of the reactance and the resistance should be the same regardless of the frequency. From (3.19) and (3.20), this requirement is expressed as

$$Q = \frac{\omega L_s}{R_s} = \frac{R_p}{\omega L_p} \tag{3.21}$$

Substituting (3.21) into (3.19) and (3.20) leads to

$$L_p = \left(1 + \frac{1}{Q^2}\right) L_s \tag{3.22}$$

$$R_p = \left(1 + Q^2\right) R_s \tag{3.23}$$

Assuming a sufficiently high Q, we can get $L_p \approx L_s$ and $R_p \approx Q^2 R_s$.

References

[1] Chattopadhyay D. *Electronics (Fundamentals and Applications)*. New York: New Age International; 2006.

[2] Nguyen NM, Meyer RG. A 1.8-GHz monolithic LC voltage-controlled oscillator. *IEEE Journal of Solid-State Circuits*. 1992;27(3):444–450.

[3] Soyuer M, Jenkins KA, Burghartz JN, Hulvey MD. A 3-V 4-GHz nMOS voltage-controlled oscillator with integrated resonator. *IEEE Journal of Solid-State Circuits*. 1996;31(12):2042–2045.

[4] Yue CP, Wong SS. On-chip spiral inductors with patterned ground shields for Si-based RF ICs. *IEEE Journal of Solid-State Circuits*. 1998;33(5): 743–752.

[5] Niknejad AM, Meyer RG. Analysis, design, and optimization of spiral inductors and transformers for Si RF ICs. *IEEE Journal of Solid-State Circuits*. 1998;33(10):1470–1481.

[6] Mohan SS, del Mar Hershenson M, Boyd SP, Lee TH. Simple accurate expressions for planar spiral inductances. *IEEE Journal of Solid-State Circuits*. 1999;34(10):1419–1424.

[7] Greenhouse HM. Design of planar rectangular microelectronic inductors. *IEEE Transactions on Parts, Hybrids, and Packaging*. 1974;10(2):101–109.

[8] Ham D, Hajimiri A. Concepts and methods in optimization of integrated LC VCOs. *IEEE Journal of Solid-State Circuits*. 2001;36(6):896–909.

[9] Mohan SS, Hershenson MD, Boyd SP, Lee TH. Bandwidth extension in CMOS with optimized on-chip inductors. *IEEE Journal of Solid-State Circuits*. 2000;35(3):346–355.

[10] Kim JK, Kim J, Kim G, Jeong DK. A Fully Integrated 0.13-μm CMOS 40-Gb/s Serial Link Transceiver. *IEEE Journal of Solid-State Circuits*. 2009;44(5):1510–1521.

[11] Huang TC, Chung TW, Chern CH, Huang MC, Lin CC, Hsueh FL. 8.4 A 28Gb/s 1pJ/b shared-inductor optical receiver with 56% chip-area reduction in 28 nm CMOS. In *2014 IEEE International Solid-State Circuits Conference Digest of Technical Papers (ISSCC)*. San Francisco, CA: IEEE; 2014 (pp. 144–145).

[12] Razavi B. *Design of Integrated Circuits for Optical Communications*. New York: John Wiley & Sons; 2012.

[13] Shekhar S, Mansuri M, O'Mahony F, *et al.* Strong injection locking in low-*Q* LC oscillators: Modeling and application in a forwarded-clock I/O receiver. *IEEE Transactions on Circuits and Systems I: Regular Papers*. 2009; 56(8):1818–1829.

[14] Nguyen NM, Meyer RG. Start-up and frequency stability in high-frequency oscillators. *IEEE Journal of Solid-State Circuits*. 1992;27(5):810–820.

[15] Bae W, Ju H, Park K, Cho SY, Jeong DK. A 7.6 mW, 414 fs RMS-jitter 10 GHz phase-locked loop for a 40 Gb/s serial link transmitter based on a two-stage ring oscillator in 65 nm CMOS. *IEEE Journal of Solid-State Circuits*. 2016;51(10):2357–2367.

[16] Rezayee A, Martin K. A 10–Gb/s clock recovery circuit with linear phase detector and coupled two–stage ring oscillator. In *Proceedings of the 28th European Solid-State Circuits Conference, 2002* (ESSCIRC 2002). Florence: IEEE; 2002 (pp. 419–422).

[17] Kwasniewski T, Abou-Seido M, Bouchet A, Gaussorgues F, Zimmerman J. Inductorless oscillator design for personal communications devices-a 1.2/spl mu/m CMOS process case study. In Pro*ceedings of the IEEE 1995 Custom Integrated Circuits Conference*. Santa Clara, CA: IEEE; 1995 (pp. 327–330).

[18] Hajimiri A, Limotyrakis S, Lee TH. Jitter and phase noise in ring oscillators. *IEEE Journal of Solid-State Circuits*. 1999;34(6):790–804.

Chapter 4
Phase noise theory for CMOS oscillators

The term "phase noise" appeared in the world in the 1950s [1–4]. During more than 60 years after the appearance, there have been tremendous number of publications related to phase noise. For example, in December 2019, Google scholar gives more than 4.5 million results in search for "phase noise"! This number may help the readers understand how important the phase noise is. In proportional to the importance, some legendary papers which provide gorgeous theory, analysis, and insight on the phase noise are ranked in the most highly cited papers in the field of circuit design. In this chapter, we will study a few selected ones from those theories to understand and to analyze the phase noise of CMOS clocking circuit.

4.1 Linear time-invariant phase noise model

Noise in oscillators can be divided into amplitude noise and phase noise. However, the amplitude fluctuations are attenuated with time by nature of amplitude restoring mechanism of an oscillator [5]. On the other hand, a phase fluctuation circulates through the oscillator forever, so that phase noise generally dominates the noise in oscillator. The Leeson model is the most well-known and intuitive model which describes the noise spectrum of an oscillator ever since it had appeared in 1966 [6]. Figure 4.1 shows the phase noise plot expected from the Leeson model. It is divided into three regions:

1. Low-frequency region that has −30 dB/dec slope, that is, the phase noise is proportional to $1/\Delta f^3$;
2. Mid-frequency region that has −20 dB/dec slope, that is, the phase noise is proportional to $1/\Delta f^2$; and
3. Flat region at high frequency.

Let us play with an example of a lossy LC tank shown in Figure 4.2 which noise processes cause the three regions. As examined in Chapter 3, assuming that the g_m is large enough to compensate the loss by R, the impedance of the tank is

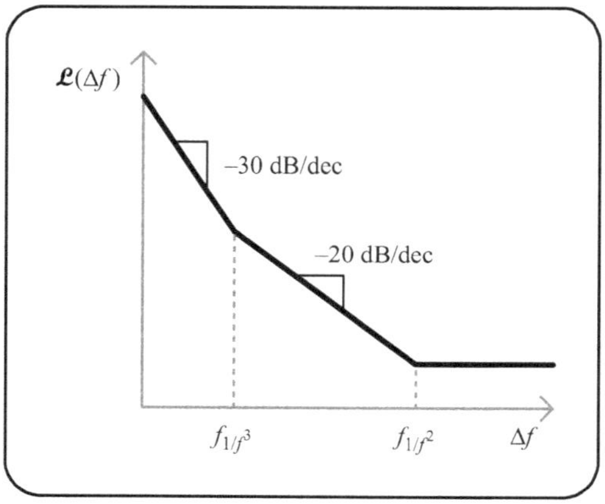

Figure 4.1 Leeson model

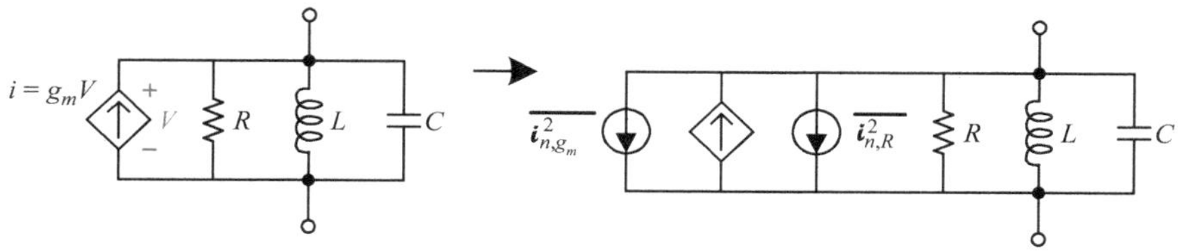

Figure 4.2 Noise sources in lossy LC tank with g_m element

infinite at the resonant frequency ω_0. Besides, from (3.1), we can calculate the impedance of the tank at the vicinity of ω_0 as

$$Z_{\mathrm{LC}(\omega_0+\Delta\omega)} = \frac{jL(\omega_0+\Delta\omega)}{1-(\omega_0+\Delta\omega)^2\mathrm{LC}} = \frac{jL(\omega_0+\Delta\omega)}{-\frac{2\Delta\omega}{\omega_0}-\frac{\Delta\omega^2}{\omega_0^2}} \tag{4.1}$$

Note that $\Delta\omega \ll \omega_0$ and $\omega_0^2 = 1/\mathrm{LC}$. Therefore, (4.1) can be approximated to

$$Z_{\mathrm{LC}(\omega_0+\Delta\omega)} = -\frac{j\omega_0 L}{\frac{2\Delta\omega}{\omega_0}} = -R\frac{\omega_0}{2Q\Delta\omega} \tag{4.2}$$

where Q is the quality factor of the inductor at the resonant frequency, which is defined as $R/\omega_0 L$. When we take account of noise currents injected from the g_m element and resistor [7], the voltage noise is expressed to

$$\overline{v_n^2}(\Delta f) = Z^2_{\mathrm{LC}(\omega_0+\Delta\omega)}\left(\overline{i^2_{n,R}} + \overline{i^2_{g_m,R}}\right) = \left(\overline{i^2_{n,R}} + \overline{i^2_{n,g_m}}\right)\left(R\frac{\omega_0}{2Q\Delta\omega}\right)^2 \tag{4.3}$$

Note that this approach assumes a linear time-invariant (LTI) system, since it does not care the dependency on the time when the noise current is injected. Also note that the oscillator is a time-variant system, so we should consider the time dependency for taking more accurate analysis. The time-variant phase noise model will be studied in Section 4.2. From the definition (2.22), the phase noise of the LC oscillator can be calculated as

$$\mathcal{L}(\Delta f) = 10\log\left(\frac{\frac{1}{2}\overline{v_n^2}(\Delta f)/\Delta f}{\frac{1}{2}V_0^2}\right) = 10\log\left(\frac{\left(\overline{i_{n,R}^2} + \overline{i_{n,g_m}^2}\right)\left(R\frac{f_0}{2Q\Delta f}\right)^2/\Delta f}{V_0^2}\right) \tag{4.4}$$

Mean square noise density of the resistor is given as

$$i_{n,R}^2/\Delta f = \frac{4kT}{R} \tag{4.5}$$

where k and T denote the Boltzmann's constant and the absolute temperature, respectively. If we assume that a MOS transistor in the saturation region is used for the g_m element, the mean square noise density is given to be

$$i_{n,g_m}^2/\Delta f = 4kT\gamma g_m\left(1 + \frac{f_{1/f}}{\Delta f}\right) \tag{4.6}$$

where γ and $f_{1/f}$ are the excess noise coefficient and the corner frequency of device $1/f$ noise, which are technology-dependent parameters. The γ is typically calculated to 2/3 for a long-channel device but increases up to 2~3 for short-channel devices. Note that the first term is the white noise component and the second term is the $1/f$ component, which is also referred to as the flicker noise. Substituting (4.5) and (4.6) into (4.4), we obtain:

$$\mathcal{L}(\Delta f) = 10\log\left(\frac{2kT(1+\gamma g_m R)\left(\frac{f_0}{2Q\Delta f}\right)^2}{\frac{V_0^2}{2R}} + \frac{2kT\gamma g_m R\left(\frac{f_0}{2Q\Delta f}\right)^2\left(\frac{f_{1/f}}{\Delta f}\right)}{\frac{V_0^2}{2R}}\right) \tag{4.7}$$

Equation (4.7) can be simplified to

$$\mathcal{L}(\Delta f) = 10\log\left(\frac{2kTF}{P_{\text{sig}}}\left(\frac{f_0}{2Q\Delta f}\right)^2 + \frac{2kT\gamma g_m R}{P_{\text{sig}}}\left(\frac{f_0}{2Q\Delta f}\right)^2\left(\frac{f_{1/f}}{\Delta f}\right)\right) \tag{4.8}$$

where $F = 1 + \gamma g_m R$, which is a factor that normalizes the total white noise to the resistor noise, and $P_{\text{sig}} = V_0^2/2R$, which represents the signal power dissipated by the resistor. Equation (4.8) implies that Q is the dominant factor for the low-noise LC oscillator. We can also see which sources cause the first and the second regions

---

$$Z_{\text{LC}} = \frac{j\omega L \cdot \frac{1}{j\omega C}}{j\omega L + \frac{1}{j\omega C}} = \frac{j\omega L}{1-\omega^2 \text{LC}} \tag{3.1}$$

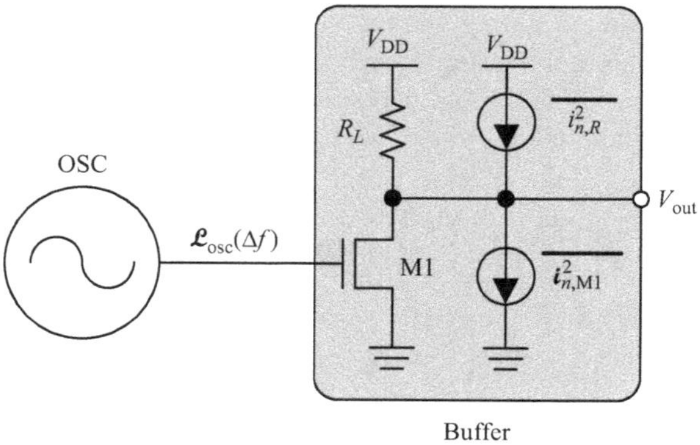

Figure 4.3 CS stage buffer driving oscillator clock

of the Leeson's model, that is, the white noise from the resistor and the g_m element causes the $1/f^2$ region, while the $1/f^3$ is caused by the $1/f$ noise. The ratio between those two noise powers is written as

$$\frac{P_{1/f}(\Delta f)}{P_{\text{white}}(\Delta f)} = \frac{\gamma g_m R}{F}\left(\frac{f_{1/f}}{f_{1/f^3}}\right) \tag{4.9}$$

At the cutoff frequency of $1/f^3$ region, (4.9) should be unity so that solving (4.9) leads to

$$f_{1/f^3} = \frac{\gamma g_m R}{1 + \gamma g_m R} \cdot f_{1/f} \tag{4.10}$$

which shows the dependency of the crossover frequency of $1/f^3$ region on the $1/f$ corner frequency of the g_m element. With $g_m R = 1$ and $\gamma > 1$, the f_{1/f^3} is larger than a half of the $f_{1/f}$, and there is no room for a designer to suppress the $1/f^3$ phase noise. Remember this result, as it will be discussed based on the time-variant phase noise model.

The last flat noise floor is induced by signal buffers. In order to measure the oscillator clock phase noise, eventually, the signal should pass any kind of several buffers. Figure 4.3 shows a simple example of a CS stage buffer. For simplicity, let us assume that the phase noise is not amplified while passing through a gain stage. Then, the buffer does not affect the $1/f^3$ and $1/f^2$ regions of the $\mathcal{L}_{osc}(\Delta f)$, and therefore V_{out} has the same phase noise curve as $\mathcal{L}_{osc}(\Delta f)$ if we neglect the noise sources at the buffer. Although the buffer noises are taken into account, those noise

---

$$\mathcal{L}(\Delta f) = 10 \log\left(\frac{P_{\text{sideband}}(f_0 + \Delta f, 1\ \text{Hz})}{P_{\text{carrier}}}\right) \tag{2.22}$$

currents are decoupled to the impedance of the LC tank. So, their contribution to the output phase noise is simply additive. As a result, assuming white noise, they introduce the white noise floor as shown in Figure 4.1.

Next, let us study another example of the CMOS ring oscillator. For simplicity, we will use the simplified first-order waveform model given in Figure 3.12. When an additive noise current i_n is injected as shown in Figure 4.4, the delay variation of each stage introduced by i_n is given as

$$\Delta\tau = \frac{\mathrm{CV_{DD}}}{2I} - \frac{\mathrm{CV_{DD}}}{2(I+i_n)} \cong \frac{\mathrm{CV_{DD}}}{2I} \cdot \frac{i_n}{I^2} \tag{4.11}$$

where the delay of a single stage τ is $\frac{\mathrm{CV_{DD}}}{2I}$. Then the frequency variation is calculated to be

$$\Delta f = \frac{1}{2N\tau} - \frac{1}{2N(\tau+\Delta\tau)} \cong \frac{1}{2N} \cdot \frac{2}{\mathrm{CV_{DD}}} \cdot i_n \tag{4.12}$$

where N is the number of stages in the ring oscillator. Then the phase modulation $\phi(t)$ in (2.11) is obtained as

$$\phi(t) = 2\pi \int_0^t \Delta f\, dx = 2\pi \cdot \frac{1}{2N} \cdot \frac{2}{\mathrm{CV_{DD}}} \cdot \int_0^t i_n dx \tag{4.13}$$

where phase is an integration of frequency. We can then calculate PSD of $\phi(t)$ as

$$S_\phi(f) = \left(2\pi \cdot \frac{1}{2N} \cdot \frac{2}{\mathrm{CV_{DD}}}\right)^2 \cdot S_W(f) \tag{4.14}$$

where

$$S_W(f) = S\left(\int_0^t i_n dx\right)(f) \tag{4.15}$$

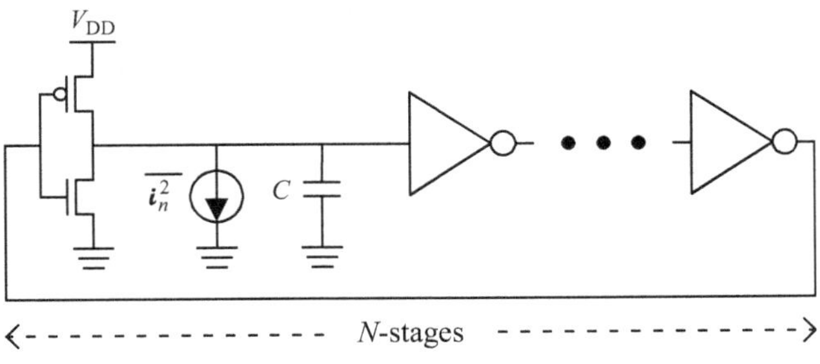

Figure 4.4 N-stage ring oscillator

$$y(t) = A_0 \sin(2\pi f_0 t + \phi_0 + \phi(t)) \tag{2.11}$$

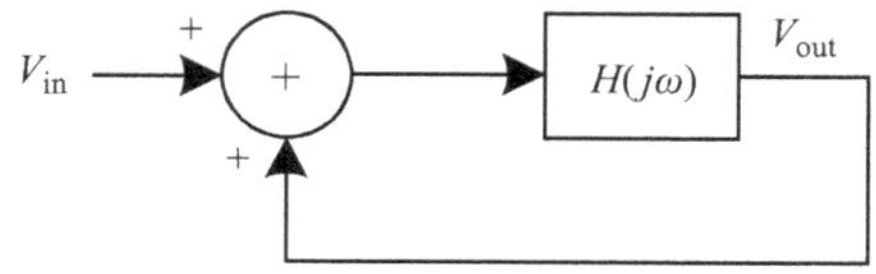

Figure 4.5 A simple feedback oscillator model

Assuming that i_n is a white noise, $S_W(f)$ becomes

$$S_W(f) = \left(\frac{1}{2\pi f}\right)^2 \cdot S_{i_n}(f) \tag{4.16}$$

where the integral of white noise is a Wiener process. Note that $S_{i_n}(f)$ equals the PSD of the white noise of the transistor, which is given as

$$\frac{i_n^2}{\Delta f} = 4kT\gamma g_m = 4kT\gamma \cdot \frac{2I}{V_{\text{DD}} - V_{\text{th}}} \tag{4.17}$$

Substituting (4.16) and (4.17), (4.14) can be written as

$$S_\phi(f) = \frac{8kT\gamma}{I(V_{\text{DD}} - V_{\text{th}})} \cdot \left(\frac{f_0}{f}\right)^2 \tag{4.18}$$

where f_0 is the oscillation frequency ($= 1/2\,N\tau$). We can also find that the white noise of the circuit element results in the $1/f^2$ of the Leeson model. The $1/f^3$ region will be discussed in Section 4.2. Equation (4.18) shows the design trade-off between the oscillation frequency and the phase noise; with the same power consumption, the lower frequency results in the better phase noise and vice versa.

In the analysis above, we verified the Leeson model for LC and ring oscillators. From now on, let us derive a bit more general interpretation of the Leeson model, which is described in [8]. For a simple feedback model of Figure 4.5, the closed-loop transfer function is given as

$$\frac{v_{\text{out}}}{v_{\text{in}}}(j\omega) = \frac{H(j\omega)}{1 - H(j\omega)} \tag{4.19}$$

At the vicinity of the oscillator frequency ω_0, the open-loop transfer function can be linearly approximated as

$$H(j(\omega_0 + \Delta\omega)) \cong H(j\omega_0) + \Delta\omega \frac{dH}{d\omega}(\omega_0) \tag{4.20}$$

Substituting (4.20) into (4.19) yields:

$$\frac{v_{\text{out}}}{v_{\text{in}}}(\omega_0 + \Delta\omega) = \frac{H(j\omega_0) + \Delta\omega \frac{dH}{d\omega}(\omega_0)}{1 - H(j\omega_0) - \Delta\omega \frac{dH}{d\omega}(\omega_0)} \tag{4.21}$$

As long as the oscillator is able to sustain oscillation, it should meet the Barkhausen criterion. Then (4.21) is reduced to

$$\frac{v_{\text{out}}}{v_{\text{in}}}(\omega_0 + \Delta\omega) = \frac{1 + \Delta\omega \frac{dH}{d\omega}(\omega_0)}{-\Delta\omega \frac{dH}{d\omega}(\omega_0)} \cong -\frac{1}{\Delta\omega \frac{dH}{d\omega}(\omega_0)} \tag{4.22}$$

Since $\frac{dH}{d\omega}(\omega_0)$ is a constant, we can find that the transfer function is proportional to $\frac{1}{\Delta\omega}$. Therefore, the PSD will be shaped by $\frac{1}{\Delta\omega^2}$, which is similar to (4.3). Therefore, considering the white noise and the flicker noise induced from the oscillator components, the output phase noise will be expressed in the combination of the $\frac{1}{\Delta\omega^2}$ and $\frac{1}{\Delta\omega^3}$ terms as (4.8), which results in the Leeson model.

4.2 Time-varying phase noise model

In Section 4.1, an oscillatory system is assumed as an LTI system, which is not true. It is a fundamentally nonlinear and time-varying system. For the linearity, however, it is reasonable to assume that the noise under consideration in the phase noise analysis is small enough that it is not worth designing or analyzing an oscillator whose noise is not small [5]. As far as the small signal is under consideration, the linearity would be a reasonable assumption. On the other hand, the phase noise of an oscillator is affected by the time of injecting noise so that the time-invariance should not be assumed for an accurate analysis of phase noise [5,9]. Prof. Thomas H. Lee at Stanford University and Prof. Ali Hajimiri at Caltech (was a graduate student at Stanford University) did several legendary contributions to establish the linear time-varying (LTV) theory on the phase noise, in the late 1990s [5,9–11]. In this chapter, we will study the LVT phase noise theory of Prof. Lee and Prof. Hajimiri.

Let us first confirm the fact that an oscillator is not a time-invariant system. Recalling the lossless LC tank in Figure 3.1(a), we can add a current source which injects a current impulse to the LC tank as shown in Figure 4.6. When the current impulse is injected at peak of the sine wave, the amplitude increases by $\Delta q/C$, where Δq is the area of the current impulse (or the amount of charge injected by the current impulse), but there is no change in phase. On the other hand, the amplitude change will persist for the lossless LC tank. However, for a practical implementation like a lossy LC tank with a g_m element given in Figure 3.4, the amplitude change will be gradually attenuated by the amplitude setting mechanism described in Figure 3.5. On the other hand, if the impulse is injected at the zero-crossing where the slope of the wave is maximized, we can find there is a change in phase. Because the oscillator does not have any phase restoring mechanism, this phase shift persists forever.

From the above observation from two extreme cases, we can find that an oscillator is time-varying system and the impulse response is a function of the time

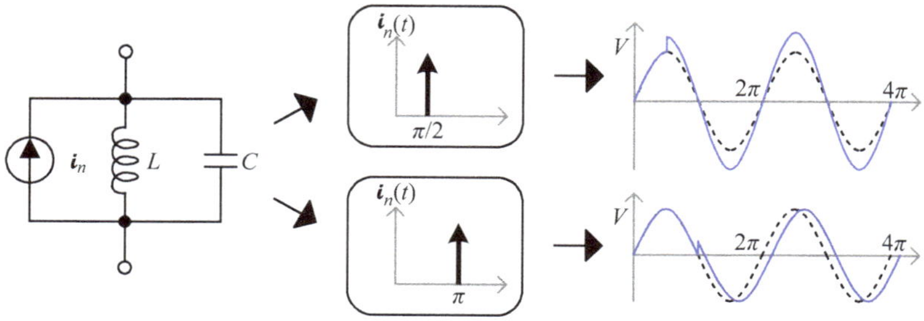

Figure 4.6 Lossless LC oscillator waveforms with respect to current pulse injection timing

of the injection. Hence, we can introduce "impulse sensitivity function (ISF)" which defines the proportionality factor between the phase shift and the voltage change by the injected charge at any point in time. Note that the ISF depends on the waveform of the signal. With ISF, we can write the phase shift as

$$\Delta\phi = \Gamma(\omega_0\tau)\frac{\Delta V}{V_{max}} = \Gamma(\omega_0\tau)\frac{\Delta q}{q_{max}} \tag{4.23}$$

where $\Gamma(\omega_0\tau)$, V_{max}, Δq, and q_{max} denote the ISF, the voltage swing across the capacitor, the amount of injected charges, and the maximum charge stored in the capacitor during oscillation, respectively. Note that the ISF is a dimensionless, frequency- and amplitude-independent function periodic in 2π.

If a continuous noise current $i(t)$ is injected into the oscillation node, Δq can be written as

$$\Delta q = i(t)\Delta t \tag{4.24}$$

Substituting (4.24) into (4.23) and integrating both sides, we can get:

$$\int_{-\infty}^{t} d\phi = \phi(t) = \int_{-\infty}^{t} \frac{\Gamma(\omega_0\tau)}{q_{max}} i(\tau)d\tau \tag{4.25}$$

As long as we consider the oscillators whose waveform should be periodic, the ISF is also periodic so that it can be expanded in a Fourier series as

$$\Gamma(\omega_0\tau) = \frac{c_0}{2} + \sum_{m=1}^{\infty} c_m \cos(m\omega_0\tau + \theta_m) \tag{4.26}$$

where c_m is the Fourier coefficient and θ_m is the phase of the mth harmonic. Combining (4.25) and (4.26), we can obtain:

$$\phi(t) = \frac{1}{q_{max}}\left(\frac{c_0}{2}\int_{-\infty}^{t} i(\tau)d\tau + \sum_{m=1}^{\infty} c_m \int_{-\infty}^{t} \cos(m\omega_0\tau + \theta_m)i(\tau)d\tau\right) \tag{4.27}$$

Note that θ_m can be neglected in the phase noise analysis where $i(t)$ is considered to be the random noise. To get more insight into physical meaning of (4.27), let us apply a slow sinusoidal input $I_0 \cos(\Delta\omega t)$ ($\Delta\omega \ll \omega_0$). Then (4.27) becomes:

$$\begin{aligned}\phi(t) &= \frac{I_0}{q_{\max}}\left(\frac{c_0}{2}\int_{-\infty}^{t}\cos(\Delta\omega\tau)d\tau + \sum_{m=1}^{\infty} c_m \int_{-\infty}^{t} \frac{\cos(m\omega_0+\Delta\omega)\tau + \cos(m\omega_0-\Delta\omega)\tau}{2}d\tau\right) \\ &= \frac{I_0}{q_{\max}}\left(\frac{c_0}{2\Delta\omega}\sin(\Delta\omega t) + \sum_{m=1}^{\infty}\frac{c_m}{2}\left(\frac{\sin((m\omega_0+\Delta\omega)t)}{m\omega_0+\Delta\omega} + \frac{\sin((m\omega_0-\Delta\omega)t)}{m\omega_0-\Delta\omega}\right)\right)\end{aligned} \tag{4.28}$$

Because $\Delta\omega \ll \omega_0$, all the harmonic terms are much smaller than the first term. Therefore, (4.28) can be simplified to

$$\phi(t) = \frac{I_0 c_0}{2q_{\max}\Delta\omega}\sin(\Delta\omega t) \tag{4.29}$$

For more general form, we can apply $I_n \cos((n\omega_0 + \Delta\omega)t)$, then (4.27) can be written as

$$\begin{aligned}\phi(t) = \frac{I_0}{q_{\max}}\Bigg(&\frac{c_0}{n\omega_0 + \Delta\omega}\sin((n\omega_0 + \Delta\omega)t) \\ &+ \sum_{m=1}^{\infty}\frac{c_m}{2}\left(\frac{\sin(((m+n)\omega_0 + \Delta\omega)t)}{(m+n)\omega_0 + \Delta\omega} + \frac{\sin(((m-n)\omega_0 - \Delta\omega)t)}{(m-n)\omega_0 - \Delta\omega}\right)\Bigg)\end{aligned} \tag{4.30}$$

In the same manner as (4.29), the last term with $m = n$ must be the dominant one. Thus (4.30) is simplified to

$$\phi(t) = \frac{I_n c_n}{2q_{\max}\Delta\omega}\sin(\Delta\omega t)\ (n > 0) \tag{4.31}$$

From (2.10), the PSD of $\phi(t)$ is calculated as

$$S_\phi(\omega) = \frac{1}{4}\left(\frac{I_n c_n}{2q_{\max}\Delta\omega}\right)^2 \delta(\omega - \Delta\omega) + \frac{1}{4}\left(\frac{I_n c_n}{2q_{\max}\Delta\omega}\right)^2 \delta(\omega + \Delta\omega) \tag{4.32}$$

The physical meaning of (4.31) and (4.32) is that the high frequency $n\omega_0 + \Delta\omega$ current disturbance is down-converted to the baseband frequency $\Delta\omega$. Therefore,

---

$$S_y(f) = \frac{1}{4}A_0^2\delta(f - f_0) + \frac{1}{4}A_0^2\delta(f + f_0) \tag{2.10}$$

when $i(t)$ with a wide spectrum is applied, the PSD of $\phi(t)$ is the sum of (4.29) and (4.31), for $n\omega_0$ in range of frequency under consideration. That is, the PSD is

$$S_\phi(\omega) = (\delta(\omega - \Delta\omega) + \delta(\omega + \Delta\omega)) \sum_{n=0}^{n_{\max}} \left(\left(\frac{I_n c_n}{4q_{\max}\Delta\omega} \right)^2 \right) \tag{4.33}$$

As defined in Chapter 2, the phase noise is achieved from $S_\phi(\omega)$ by converting to single-sideband and having logarithmic expression:

$$\mathcal{L}(\Delta\omega) = 10 \log \left(\frac{\sum_{n=0}^{n_{\max}} (I_n^2 c_n^2)}{8q_{\max}^2 \Delta\omega^2} \right) \tag{4.34}$$

If we assume that the noise is a white noise, then the peak amplitude I_n is replaced by the noise spectral density as

$$\frac{I_0^2}{2} = \frac{I_1^2}{2} = \ldots = \frac{I_n^2}{2} = \frac{\overline{i_n^2}}{\Delta f} \tag{4.35}$$

Equation (4.34) is simplified to

$$\mathcal{L}(\Delta\omega) = 10 \log \left(\frac{\frac{\overline{i_n^2}}{\Delta f} \sum_{n=0}^{\infty} c_n^2}{4q_{\max}^2 \Delta\omega^2} \right) \tag{4.36}$$

According to Parseval's theorem, the sum of c_n^2 is expressed with the ISF as

$$\sum_{n=0}^{\infty} c_n^2 \approx \frac{c_0^2}{2} + \sum_{n=1}^{\infty} c_n^2 = \frac{1}{\pi} \int_{-\pi}^{\pi} |\Gamma(x)|^2 dx \tag{4.37}$$

where c_0 is assumed to be small, which is a valid assumption when the ISF is symmetrical. We finally get the LTV phase noise model due to the white noise as

$$\mathcal{L}(\Delta\omega) = 10 \log \left(\frac{\Gamma_{\text{rms}}^2}{2q_{\max}^2} \cdot \frac{\frac{\overline{i_n^2}}{\Delta f}}{\Delta\omega^2} \right) \tag{4.38}$$

The $1/f^3$ region of the Leeson Model can be also achieved from (4.34). As expected from the LTI derivation (4.8) and (4.10), the $1/f^3$ region is induced from the $1/f$ noise which is expressed as

$$\overline{i_{n,1/f}^2} = \overline{i_n^2} \cdot \frac{\omega_{1/f}}{\Delta\omega} \tag{4.39}$$

Unless the oscillator frequency is extremely slow, $\omega_{1/f}$ is much lower than ω_0 so that only the case of $n = 0$ is considered. As a result, (4.34) is simplified to

$$\mathcal{L}(\Delta\omega) = 10 \log \left(\frac{c_0^2}{4q_{\max}^2} \cdot \frac{\frac{\overline{i_n^2}}{\Delta f}}{\Delta\omega^2} \cdot \frac{\omega_{1/f}}{\Delta\omega} \right) \tag{4.40}$$

The $1/f^3$ cutoff frequency $\Delta\omega_{1/f^3}$ is achieved by equating (4.38) and (4.40)

$$10\log\left(\frac{\Gamma_{\mathrm{rms}}^2}{2q_{\max}^2}\cdot\frac{\frac{\overline{i_n^2}}{\Delta f}}{\Delta\omega_{1/f^3}{}^2}\right)=10\log\left(\frac{c_0^2}{4q_{\max}^2}\cdot\frac{\frac{\overline{i_n^2}}{\Delta f}}{\Delta\omega_{1/f^3}{}^2}\cdot\frac{\omega_{1/f}}{\Delta\omega_{1/f^3}}\right) \tag{4.41}$$

which leads to

$$\Delta\omega_{1/f^3}=\omega_{1/f}\cdot\frac{c_0^2}{2\Gamma_{\mathrm{rms}}^2} \tag{4.42}$$

From the definition, c_0 is calculated as

$$c_0=\frac{1}{\pi}\int_{-\pi}^{\pi}\Gamma(x)dx=2\Gamma_{\mathrm{dc}} \tag{4.43}$$

where Γ_{dc} is the DC component of the ISF. We can find that $\Delta\omega_{1/f^3}$ is scaled by $\frac{c_0^2}{2\Gamma_{\mathrm{rms}}^2}$, hence reducing the DC component of the ISF can significantly suppress $1/f^3$ phase noise. This insight cannot be achieved from LTI derivation of (4.10), showing the powerfulness of the LTV analysis. Definitely, (4.38) and (4.42) imply that achieving the ISF function is the most important task to calculate the phase noise. From now, we will derive the ISF of an LC oscillator and a ring oscillator, which are the examples we already investigated using the LTI analysis, so that we can compare the LTI and LTV analyses. For LC oscillator shown in Figure 4.7, once again it is assumed that the g_m fully compensates the resistor loss for simplicity. The voltage across the capacitor and the current through the inductor can be expressed as

$$V_C(t)=V_0\cos(\omega t) \tag{4.44}$$

$$I_L(t)=V_0\sqrt{\frac{C}{L}}\sin(\omega t) \tag{4.45}$$

When the current impulse injects a small amount of charge of Δq at $t=\tau$, the voltage instantly increases by $\Delta q/C$ but the current stays. This instant event will be

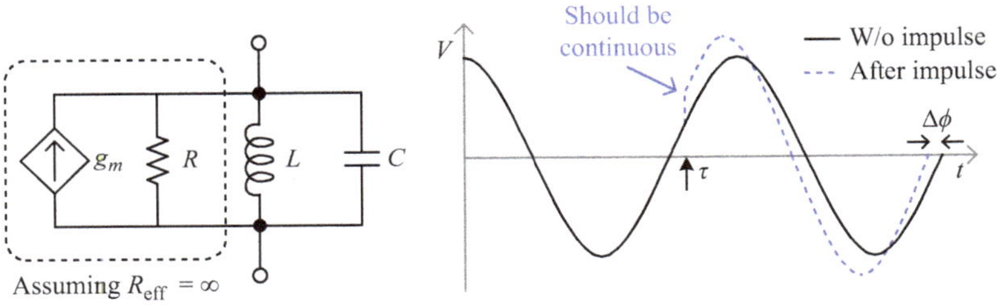

Figure 4.7 ISF of LC oscillator

converted to a small change in both the voltage swing (ΔV) and phase ($\Delta\phi$) of the oscillation signal, as we observed in Figure 4.6. The voltage and current after the injection can be written as

$$\begin{aligned} V_C(t) &= (V_0 + \Delta V)\cos(\omega t + \Delta\phi) \\ &= (V_0 + \Delta V)(\cos(\omega t)\cos(\Delta\phi) - \sin(\omega t)\sin(\Delta\phi)) \end{aligned} \tag{4.46}$$

$$\begin{aligned} I_L(t) &= (V_0 + \Delta V)\sqrt{\frac{C}{L}}\sin(\omega t + \Delta\phi) \\ &= (V_0 + \Delta V)\sqrt{\frac{C}{L}}(\cos(\omega t)\cos(\Delta\phi) + \sin(\omega t)\sin(\Delta\phi)) \end{aligned} \tag{4.47}$$

Note that the voltage and current should be continuous at $t = \tau$. Then we have simultaneous equations as

$$V_0\cos(\omega\tau) + \frac{\Delta q}{C} = (V_0 + \Delta V)(\cos(\omega t)\cos(\Delta\phi) - \sin(\omega t)\sin(\Delta\phi)) \tag{4.48}$$

$$V_0\sqrt{\frac{C}{L}}\sin(\omega\tau) = (V_0 + \Delta V)\sqrt{\frac{C}{L}}(\cos(\omega t)\cos(\Delta\phi) + \sin(\omega t)\sin(\Delta\phi)) \tag{4.49}$$

For small enough ΔV and $\Delta\phi$, we can simplify (4.48) and (4.49) using approximations of $\cos(\Delta\phi) \approx 1$, $\sin(\Delta\phi) \approx \Delta\phi$, and $V_0 + \Delta V \approx V_0$ as

$$\Delta V\cos(\omega\tau) - V_0\,\Delta\phi\,\sin(\omega\tau) = \frac{\Delta q}{C} \tag{4.50}$$

$$\Delta V\sin(\omega\tau) - V_0\,\Delta\phi\,\cos(\omega\tau) = 0 \tag{4.51}$$

By solving (4.50) and (4.51), the relation between $\Delta\phi$ and Δq can be calculated to

$$\Delta\phi = -\frac{\Delta q}{\mathrm{CV}_0}\sin(\omega\tau) \tag{4.52}$$

From (4.23), we can obtain the ISF of the LC oscillator as

$$\Gamma(\omega_0\tau) = -\sin(\omega\tau) \tag{4.53}$$

Substituting (4.53) and the white noise of (4.5) and (4.6) into (4.38), the phase noise of LC oscillator is calculated as

$$\begin{aligned} \mathcal{L}(\Delta\omega) &= 10\log\left(\frac{kT}{q_{\max}^2}\cdot\frac{\frac{1}{R} + \gamma g_m}{\Delta\omega^2}\right) = 10\log\left(\frac{kT}{C^2V_{\text{swing}}^2}\cdot\frac{\frac{F}{R}}{\Delta\omega^2}\right) \\ &= 10\log\left(\frac{kTF}{2R^2C^2P_{\text{sig}}}\cdot\frac{1}{\Delta\omega^2}\right) \end{aligned} \tag{4.54}$$

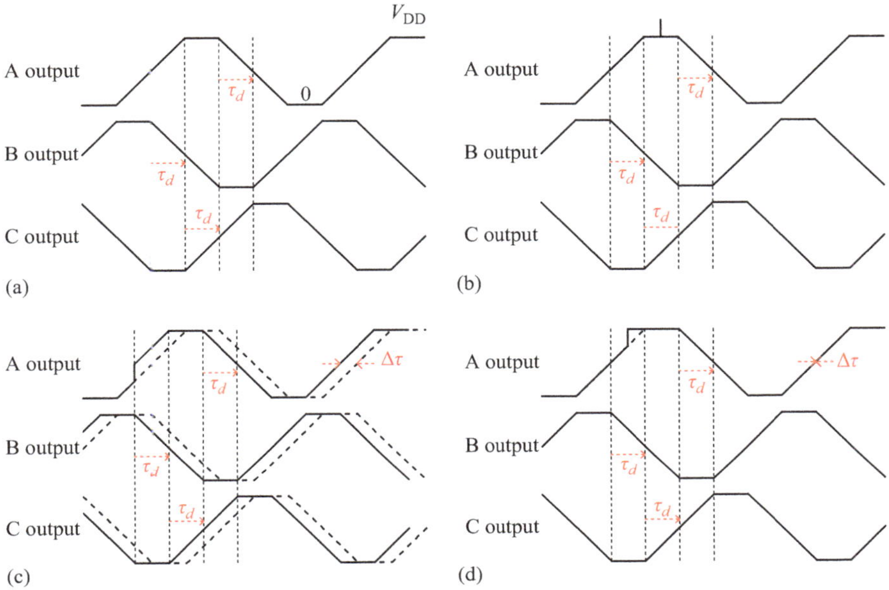

Figure 4.8 Case study of the charge injection to ring oscillator: (a) no charge injection, (b) charge injection when the voltage is settled to V_{DD}/GND, (c) charge injection in the first half of the transition, and (d) charge injection in the second half of the transition

where P_{sig} is the power dissipation of the LC oscillator, which is calculated here as ${V_{swing}}^2/2R$. Substituting $R^2 = Q^2L/C$, we can find that (4.54) is actually the same as what we obtained from the LTV analysis in (4.8):

The next example we will investigate is the ring oscillator. For simplicity, the three-stage ring oscillator (refer to Figure 3.12) is used, and the first-order waveform model shown in Figure 3.12 is assumed. That is, NMOS and PMOS of each inverter turn on when the gate-source voltage is higher than $V_{DD}/2$ and flow a uniform current while turning on. Similar to the LC example, it is useful to examine the tendency of the phase shift due to the charge injection, with respect to the impulse timing. Figure 4.8 shows three cases of the impulse timing, where τ_d is the time delay of each inverter. When the voltage is settled to V_{DD}, the PMOS of inverter is turning on so the injected charge instantly flows out through the PMOS. As a result, such charge injection does not introduce the phase shift at all. The case of the injection during transition is divided into two cases. If the impulse occurs during the first half of the transition, the injected charges accelerate the transition

---

$$\Delta\phi = \Gamma(\omega_0\tau)\frac{\Delta V}{V_{max}} = \Gamma(\omega_0\tau)\frac{\Delta q}{q_{max}} \tag{4.23}$$

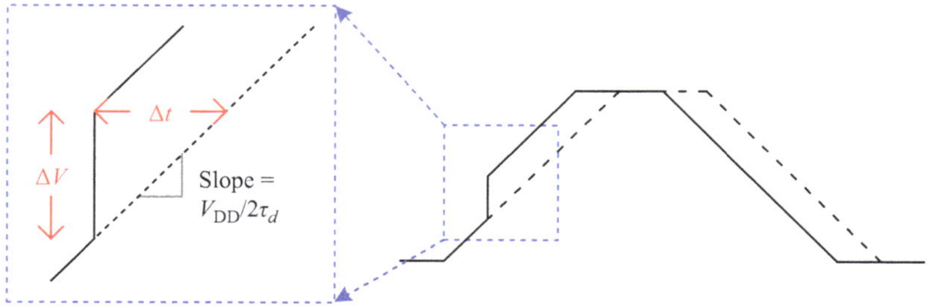

Figure 4.9 Relation between the phase shift and the slope of transition

time to $V_{DD}/2$, the delay of the inverter stage. Consequently, the second inverter is triggered earlier, and the induced phase shift sequentially propagates through the ring and turns back to the first inverter. In other words, such injection introduces phase shift, which circulates through the ring permanently. On the other hand, in the second half of the transition, the voltage is already higher than $V_{DD}/2$ so that the charge injection does not affect the transition of the second inverter. It may seem to be weird, but it is because we simplified the nonlinear operation of ring oscillator with the first-order transition model which does not reflect a real circuit operation. Note that if we adopt a higher-order model the calculation will be complex.

Figure 4.9 helps a quantitative calculation of the ISF analysis. Since the rise/fall time of the first-order waveform equals $2\tau_d$ while the voltage swing equals V_{DD}, the slope of transition is $V_{DD}/2\tau_d$. Therefore, the amount of phase shift can be written as

$$\Delta\phi = \frac{2\pi}{T_0}\cdot\frac{2\tau_d\Delta V}{V_{DD}} = 4\pi f_0\tau_d\cdot\frac{\Delta q}{CV_{DD}} \tag{4.55}$$

where C denotes the total load capacitance driven by each inverter stage. From (4.23) and $q_{max} = CV_{DD}$, the ISF of the CMOS ring oscillator is obtained, as shown in Figure 4.10. Then the Γ^2_{rms} is calculated as

$$\Gamma^2_{rms} = \frac{2\cdot 16\pi^2 f_0^2\tau_d^2}{2Nf_0}\cdot f_0 = \frac{16\pi^2 f_0^2\tau_d^2}{N} \tag{4.56}$$

Note that this result comes from only one of the oscillation nodes in the ring, but there are N nodes in the ring oscillator. Assuming that the noise sources of each stage are not correlated with each other, (4.38) becomes:

$$\mathcal{L}(\Delta\omega) = 10\log\left(N\cdot\frac{16\pi^2 f_0^2\tau_d^2}{N}\cdot\frac{1}{2q_{max}^2}\cdot\frac{\frac{\overline{i_n^2}}{\Delta f}}{\Delta\omega^2}\right) \tag{4.57}$$

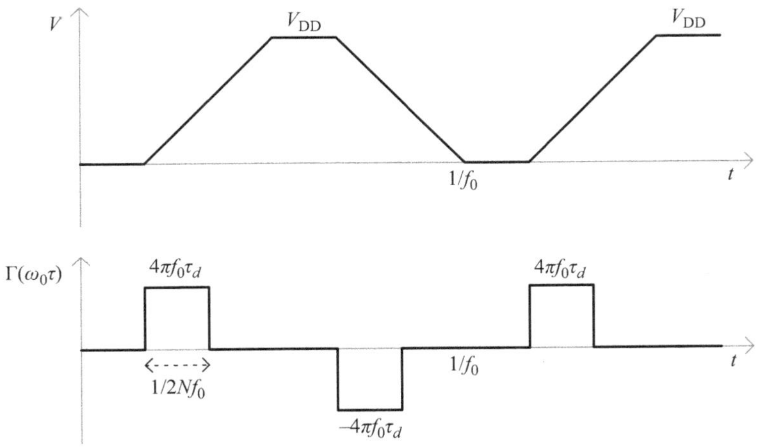

Figure 4.10 ISF of the CMOS ring oscillator

By substituting $q_{max} = CV_{DD}$ and:

$$\tau_d = \frac{CV_{DD}}{I} \tag{4.58}$$

Equation (4.57) can be simplified to

$$\mathcal{L}(\Delta\omega) = 10\log\left(\frac{8\pi^2 f_0^2}{I^2}\cdot\frac{\frac{\overline{i_n^2}}{\Delta f}}{\Delta\omega^2}\right) = 10\log\left(\frac{2}{I^2}\cdot\frac{\omega_0^2}{\Delta\omega^2}\cdot\frac{\overline{i_n^2}}{\Delta f}\right) \tag{4.59}$$

Assuming that g_m of PMOS and NMOS are equal, combining the white noise component of the transistor noise leads to

$$\mathcal{L}(\Delta\omega) = 10\log\left(\frac{8kT\gamma}{I}\cdot\frac{g_m}{I}\cdot\frac{\omega_0^2}{\Delta\omega^2}\right) = 10\log\left(\frac{8kT\gamma}{I}\cdot\frac{2}{V_{OV}}\cdot\frac{\omega_0^2}{\Delta\omega^2}\right) \tag{4.60}$$

where V_{OV} denotes to the gate-overdrive voltage, $V_{GS}-V_{TH}$. We can find that we obtained the same result from the LTV analysis in (4.18) (do not forget DSB-SSB conversion). Equation (4.60) lets us know the important criterion for the ring oscillator design. There are three variables: the current consumption, the gate-overdrive voltage, and the oscillation frequency. Note that (4.60) is not a function of the number of stages. The phase noise increases quadratically as the oscillation frequency increases if the other conditions stay the same. The current consumption and the gate-overdrive voltage are inversely proportional to the phase noise. Since the gate-overdrive voltage of CMOS inverter increases as the supply voltage increases or the threshold voltage decreases, which are generally more like the CMOS technology parameters rather than the design

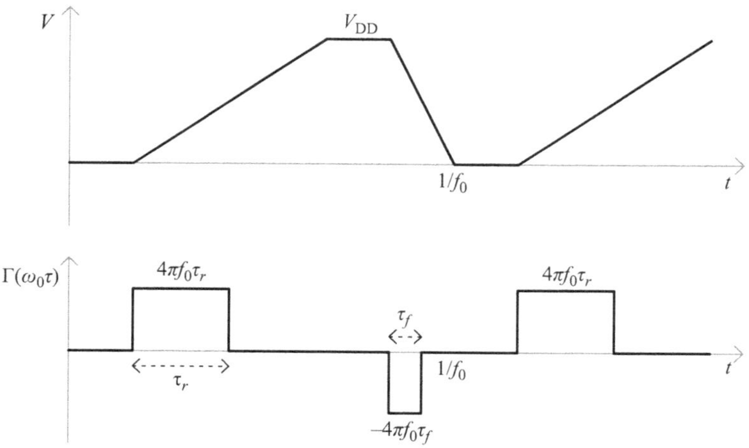

Figure 4.11 Waveform of ring oscillator when NMOS is stronger than PMOS

parameters, the dependency on the gate-overdrive voltage is not significant or controllable. As a result, the phase noise in the $1/f^2$ region is a strong function of the current consumption of the ring oscillator as long as the oscillation frequency is fixed.

As observed from (4.42) and (4.43), the symmetric waveform like Figure 4.10 suppresses the $1/f$ noise conversion to $1/f^3$ phase noise. However, due to the inherent mismatch between NMOS and PMOS, the waveform can never be symmetric, therefore it should be useful to study $1/f^3$ phase noise caused by the asymmetric waveform. Assuming the rise delay (τ_r) and the fall delay (τ_f) are not matched, the waveform and the ISF become as shown in Figure 4.11. The Γ_{rms}^2 and Γ_{DC} can be calculated as

$$\Gamma_{\text{rms}}^2 = 16\pi^2 f_0^2\left(\tau_r^3 + \tau_f^3\right)\cdot f_0 = 16\pi^2 f_0^3\tau_r^3\left(1+\beta^3\right) \tag{4.61}$$

$$\Gamma_{\text{DC}} = 4\pi f_0\left(\tau_r^2 - \tau_f^2\right)\cdot f_0 = 4\pi f_0^2\tau_r^2\left(1-\beta^2\right) \tag{4.62}$$

where β is the ratio of the fall delay over the rise delay. Combining (4.61) and (4.62), (4.42) becomes:

$$\Delta\omega_{1/f^3} = \omega_{1/f}\cdot\frac{2f_0\tau_r\left(1-\beta^2\right)^2}{\left(1+\beta^3\right)} \tag{4.63}$$

By substituting:

$$f_0 = \frac{1}{N\left(\tau_r+\tau_f\right)} = \frac{1}{N\tau_r(1+\beta)} \tag{4.64}$$

Equation (4.63) is simplified to

$$\Delta\omega_{1/f^3} = \omega_{1/f} \cdot \frac{2(1-\beta^2)^2}{N(1+\beta)(1+\beta^3)} = \omega_{1/f} \cdot \frac{2(1-\beta)^2}{N(1-\beta+\beta^2)} \tag{4.65}$$

which implies that the rise delay and fall delay should be matched to suppress the $1/f^3$ noise. If they are not matched, the $1/f^3$ corner frequency is inversely proportional to the number of stages. Note that the $1/f^2$ phase noise was not a function of N. This observation may mislead to a wrong conclusion that higher N results in a lower phase noise. However, for a fixed frequency, increasing N is equivalent to increasing the power consumption. Combined (4.58) and (4.64), (4.65) can be rewritten as

$$\Delta\omega_{1/f^3} = \omega_{1/f} \cdot \frac{2(1-\beta^2)^2}{N(1+\beta)(1+\beta^3)} = \omega_{1/f} \cdot \frac{2(1-\beta)^2}{(1-\beta+\beta^2)} \cdot \frac{I}{\mathrm{C}V_{\mathrm{DD}}f_0} \tag{4.66}$$

For the same power consumption, supply voltage, and frequency, the $1/f^3$ does not depend on the N as well. Sometimes, the power consumption increases super-linearly as N increases at the same oscillator frequency, and therefore using a larger device for smaller N is probably a better solution because it reduces the $\omega_{1/f}$ itself.

The time-varying model is also useful to evaluate the noise contribution from the cross-coupled inverters in a differential ring oscillator. Symbol and schematic diagrams of a pseudo-differential ring oscillator are shown in Figure 4.12(a). With the LTI theory where we do not consider the time dependency of noise injection, the total NMOS current noise injected to the node O− is expressed with the sum of transconductances of MN1 (main buffer) and LN1 (cross-coupled latch) as

$$i^2_{n,\mathrm{total}} = 4kT\gamma\left(g_{m,\mathrm{MN1}} + g_{m,\mathrm{LN1}}\right) \tag{4.67}$$

Note that only NMOS devices are considered for simplicity. Given the latch ratio α, which represents the ratio of driving strength of the cross-coupled inverter over the main buffer, (4.67) is approximated as

$$i^2_{n,\mathrm{total}} = 4kT\gamma g_{m,\mathrm{MN1}}(1+\alpha) \tag{4.68}$$

Equation (4.68) implies that the MN1 and LN1 equally contribute to the total noise in proportional to their driving strength, which is a strong function of their size ratio in general. However, in reality, the noise contribution by the cross-coupled latch is much less than that from the main buffer [12]. Figure 4.12(b) gives a qualitative analysis on the noise contribution of a differential ring oscillator. In fact, there are two main factors that suppress the noise from the latch. At first, the latch devices are not turned on as strong as the main devices during the normal operation, so their transconductance is less than the main buffer's transconductance. Simplified transient waveforms of input/output voltages of a differential buffer in a ring oscillator are shown in Figure 4.12(b), where it is assumed that the transient waveforms are mainly set by the main buffers. For the MN1 which is the pull-down device of the

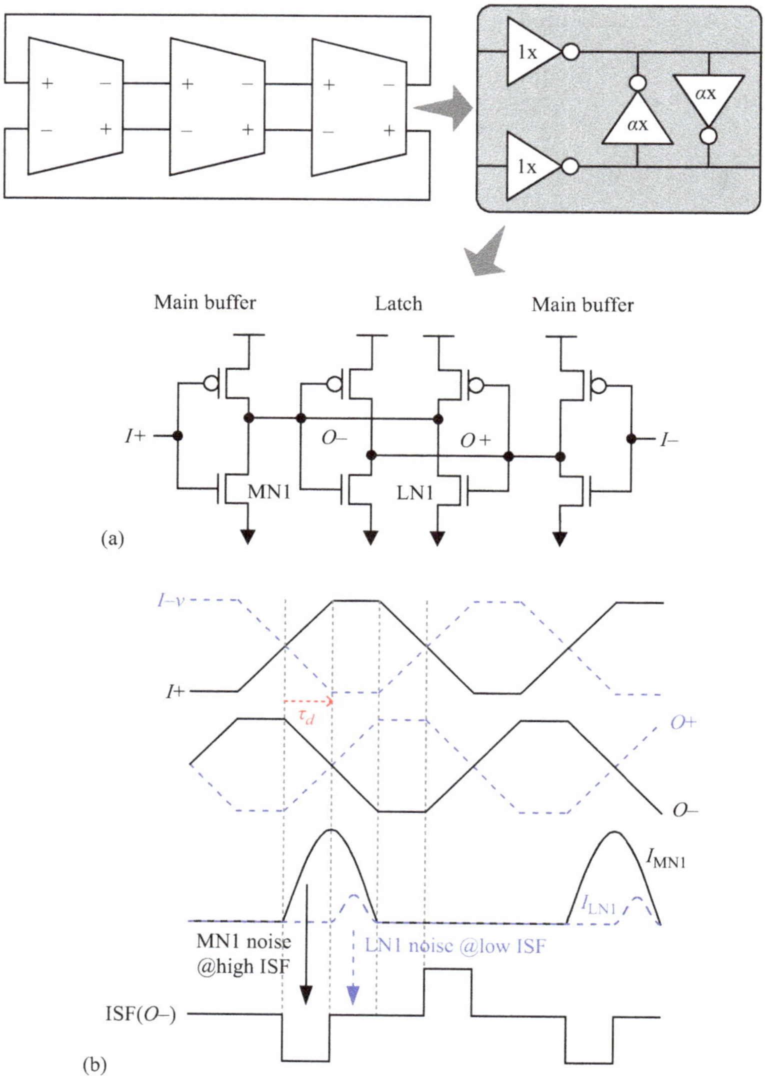

Figure 4.12 (a) Circuit diagram of a pseudo-differential CMOS ring oscillator and (b) timing diagram of nodal voltages, currents of NMOS transistors in the ring oscillator, and ISF

main buffer, the current is set by the nodal voltages at I+ (V_{GS}) and O− (V_{DS}). From the waveform, we can find that the V_{GS} exceeds $V_{DD}/2$ when the V_{DS} is sufficiently high so that the MN1 is able to flow a full saturation current. On the other hand, the gate and drain voltages of the LN1 (pull-down device of the latch) are O+ and O−, respectively, which are fully differential. As a result, the V_{GS} and the V_{DS} switch in a complementary manner so that the gate of the LN1 is turned on when the V_{DS} is low, and therefore the LN1 is not able to flow a full current. To summarize, the g_m of LN1 is less than that of MN1 because it cannot be fully turned on during normal operation.

Second, the noise of the LN1 is injected only when the ISF is minimized. From the observation from Figure 4.8, we found that the ISF has its maximum value during the first half of the transition. On the other hand, the LN1 is turned on only during the second half of the transition in contrast to the MN1 which is responsible for the transition and therefore is turned on for the entire transition. As a result, the noise current from LN1 flows only when the ISF is minimized so that the phase noise conversion is considerably suppressed. In [12], it is demonstrated that the phase noise contribution of the latch is more than 12 dB less than that of the main buffer even if the latch size is considerably large.

References

[1] Jaffe R, Rechtin E. Design and performance of phase-lock circuits capable of near-optimum performance over a wide range of input signal and noise levels. *IRE Transactions on Information Theory*. 1955;1(1):66–76.

[2] Bennett WR. Methods of solving noise problems. *Proceedings of the IRE*. 1956;44(5):609–638.

[3] Jacobsen BB. Thermal noise in multi-section radio links. *Proceedings of the IEE-Part C: Monographs*. 1958;105(7):139–150.

[4] Leeson DB. Oscillator phase noise: a 50-year review. *IEEE Transactions on Ultrasonics, Ferroelectrics, and Frequency Control*. 2016;63(8):1208–1225.

[5] Lee TH. Oscillator phase noise: A tutorial. In *Custom Integrated Circuits, 1999. Proceedings of the IEEE 1999*. San Diego, CA: IEEE; 1999 (pp. 373–380).

[6] Lesson DB. A simple model of feedback oscillator noise spectrum. *Proceedings of the IEEE*. 1966;54(2):329–330.

[7] Enz C, Chicco F, Pezzotta A. Nanoscale MOSFET modeling: Part 2: Using the inversion coefficient as the primary design parameter. *IEEE Solid-State Circuits Magazine*. 2017;9(4):73–81.

[8] Razavi B. A study of phase noise in CMOS oscillators. *IEEE Journal of Solid-State Circuits*. 1996;31(3):331–343.

[9] Hajimiri A, Lee TH. A general theory of phase noise in electrical oscillators. *IEEE Journal of Solid-State Circuits*. 1998;33(2):179–194.

[10] Hajimiri A, Limotyrakis S, Lee TH. Jitter and phase noise in ring oscillators. *IEEE Journal of Solid-State Circuits*. 1999;34(6):790–804.

[11] Lee TH, Hajimiri A. Oscillator phase noise: A tutorial. *IEEE Journal of Solid-State Circuits*. 2000;35(3):326–336.

[12] Bae W, Ju H, Park K, Cho SY, Jeong DK. A 7.6 mW, 414 fs RMS-jitter 10 GHz phase-locked loop for a 40 Gb/s serial link transmitter based on a two-stage ring oscillator in 65 nm CMOS. *IEEE Journal of Solid-State Circuits*. 2016;51(10):2357–2367.

Chapter 5
Introduction to PLL/DLL

In Chapter 4, we studied the phase noise of electrical oscillators. As described in Chapter 4, the frequency instability of an oscillator diverges at a low frequency. It limits the use of the oscillator standalone as a clocking circuit in many applications. To address that, a negative feedback system is generally utilized to correct the instability of the oscillator. A phase-locked loop (PLL) forms a negative feedback loop where an oscillator-generated signal is frequency- and phase-locked to a reference signal. Just like any feedback loop, a PLL comprises of producer, sensor, and loop filter. As shown in the simplified block diagram of PLL in Figure 5.1, the voltage-controlled oscillator (VCO) which generates a clock whose frequency is controlled by a voltage input serves as the producer, and the phase detector which measures the phase difference between the VCO clock and the reference clock is used for the sensor of the loop. The loop filter determines how to control the producer based on the measured value from the sensor. Because a PLL controls the frequency of the VCO to match the phase and frequency of the reference clock and the output clock, the loop filter needs to adjust both the phase and frequency so that the system should be second-order or higher. On the other hand, a delay-locked loop (DLL) uses a voltage-controlled delay line (VCDL) as the producer. Since the VCDL receives an input signal and adjust only the delay, the frequency of the reference clock and the output clock is always the same, whereas a PLL adjusts the frequency to correct the phase error. Therefore, a DLL does not have to be a second-order system. For well-designed PLLs/DLLs, the phase error is gradually corrected by the negative feedback and eventually converges to zero.

5.1 Applications of PLL/DLL

Due to the importance of a precise clock in electronic systems, PLLs/DLLs have been almost ubiquitous in a variety of integrated circuits (ICs). A few selected examples of PLL/DLL applications will be introduced in this chapter. One of the most important applications of the PLL is clock frequency multiplier. Although crystal oscillators provide a low-noise clock due to their extremely high Q and excellent temperature stability, their low frequency which is typically lower than 100 MHz restricts their usage in modern ICs [1]. A PLL configuration shown in Figure 5.2(a) which places a frequency divider in the feedback path can be used to multiplying the reference frequency provided from a crystal oscillator since the

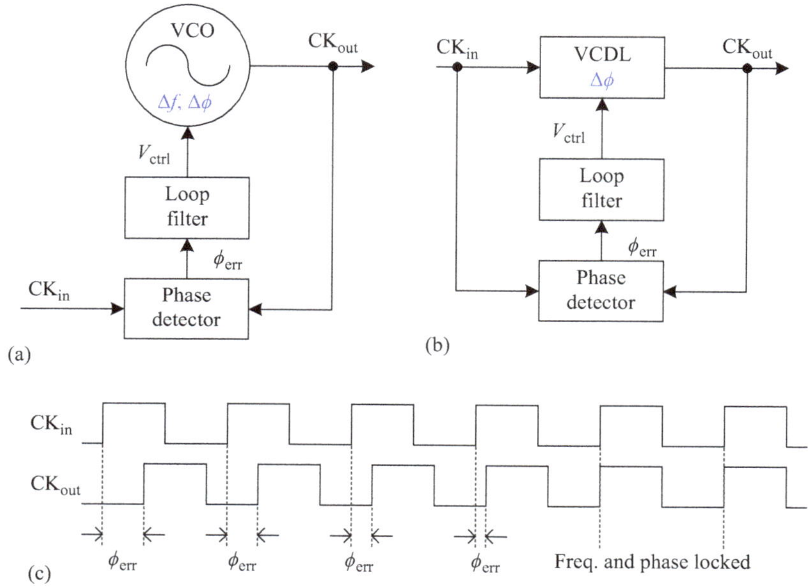

Figure 5.1 (a) Simplified block diagram of PLL, (b) DLL, and (c) and locking transient of PLL/DLL

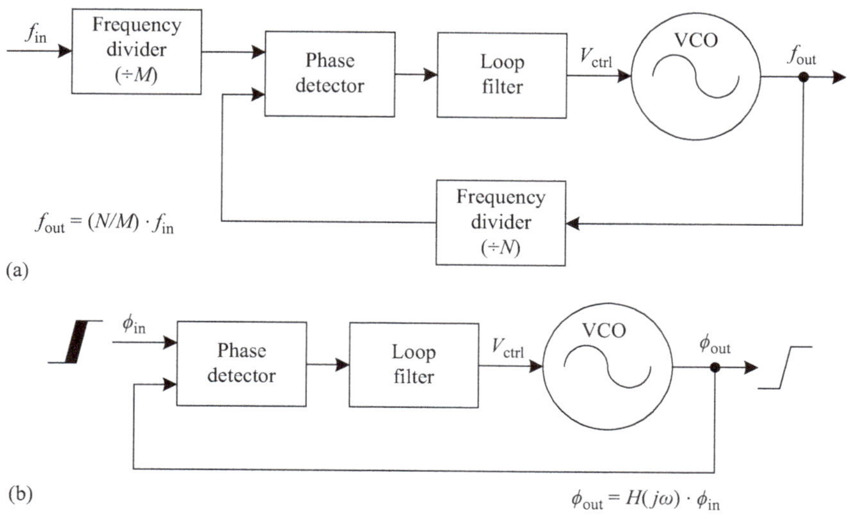

Figure 5.2 PLL application as (a) clock multiplier and (b) jitter filter

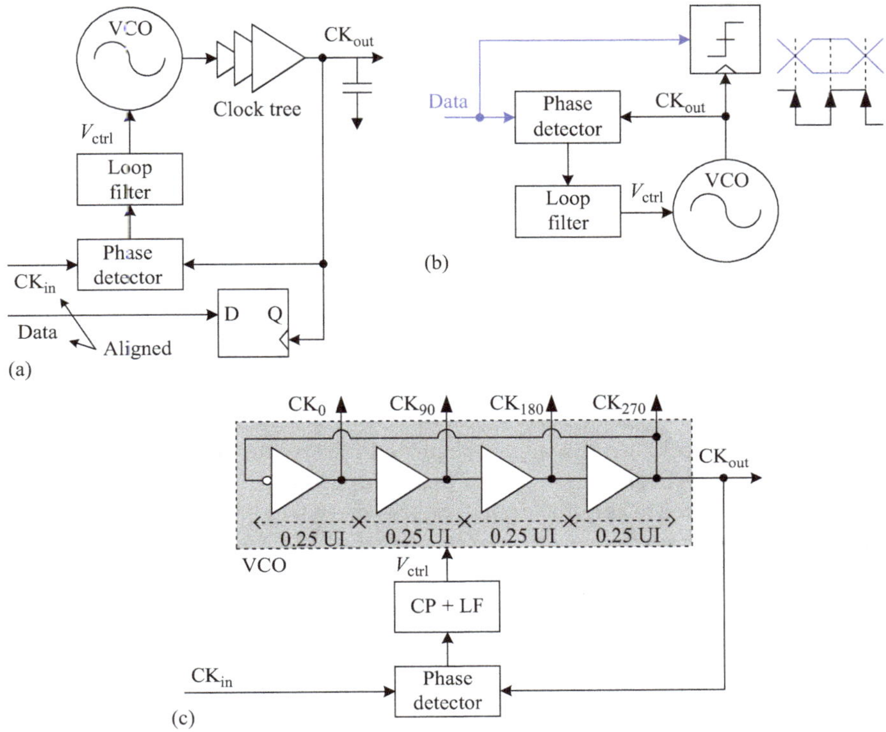

Figure 5.3 PLL used as (a) zero delay buffer, (b) clock and data recovery circuit, and (c) multiphase clock generator

PLL locks the frequency and phase of the input clock and the feedback clock. Using another frequency divider at the input, a fractional multiplication ratio can be achieved. As will be studied in Chapter 6, reducing the input reference frequency may degrade the jitter performance of the PLL, so that the frequency divider ($1/N$) at the feedback path is generally replaced by a fractional divider (M/N) to avoid dividing the reference frequency. A fractional-N clock multiplier will be introduced in Section 5.3. For DLL, the frequency multiplication is possible but not as straightforward as for the PLL. The clock multiplication technique using DLL will be studied in Chapter 11.

Another example of PLL applications is a jitter filter. As will be studied in Chapter 6, the PLL exhibits a low-pass transfer function for the phase noise from the input. As a result, the input jitter component whose frequency is higher than the PLL loop bandwidth is filtered out at the output [2]. For DLL, the jitter filtering capability depends on the DLL topology; however, a general DLL is not able to filter out the input jitter. The DLL jitter characteristic will be covered in Chapter 7.

Figure 5.3 shows three other PLL applications. The first one is a zero-delay buffer, which sounds physically impossible. Either of inter-chip and chip-to-chip communication, a clock signal typically has a large load so that a buffer chain or a clock tree is usually used to drive and distribute the clock to the load. Because a buffer stage introduces a nonzero delay to the signal path, a significant skew is added between the clock and data. The amount of skew depends on the clock loading, the process technology, and the process, voltage, and temperature (PVT) variations, which lead to a considerable uncertainty on the timing, even though the clock and data are initially synchronized [3]. By placing the clock tree or the clock buffer in the feedback loop, as shown in Figure 5.3(a), the phase lock as well as the clock driving strength are achieved simultaneously. Moreover, the PLL can also suppress the input jitter utilizing the jitter filtering capability mentioned above. A similar timing issue arises in a serial link transceiver which is widely used for chip-to-chip communication links. The timing issue is even more severe because the SNR is limited in this application so that a slight skew results in a significant degradation on the link bit-error-rate (BER) [4,5]. Using the nonreturn-to-zero (NRZ) data for the input of the PLL instead of a clock, the PLL aligns the phase of the output clock to the NRZ data transition, so that the data are sampled at the optimal timing. This kind of configurations is widely referred to as a clock and data recovery (CDR) circuit. Multiphase clock generation is another very important application of the PLL. As shown in Figure 5.3(c), a PLL based on an N-stage ring oscillator generates evenly spaced N clock phases. In a sense, the multiphase generation is one of the most frequently used applications of PLL because it can be combined with most of the other applications of PLL. After all, it can be used in many signal processing paths where a time-interleaving technique is adopted to achieve high aggregate speed [6], such as high-speed I/O circuits and radio frequency (RF) communication circuits [7,8]. Figure 5.4 shows DLL-based implementations of the applications mentioned above. Note that the VCDL does not produce the clock by itself, so a clock input should be provided to the VCDL.

5.2 Building blocks

5.2.1 Voltage-controlled oscillator

An oscillator whose frequency is adjusted in accordance with the input voltage is referred to as VCO. Recalling Chapters 3 and 4, the LC oscillator and the ring oscillator are the most frequently used oscillator topology, and the oscillation frequency and the frequency stability are the important metrics to evaluate an oscillator. In addition, a tuning range and a VCO gain (K_{VCO}) which mean the frequency range available in the given input voltage range and the slope of VCO curve, respectively, are another important parameters for a VCO (Figure 5.5). When we compare the LC-VCO (Figure 5.6) and ring-VCO (Figure 5.7) in a general manner, the LC-VCO has advantages of a lower phase noise, a less sensitivity to the PVT variations, and a higher frequency; on the other hand, the ring-VCO exhibits a wider tuning range, a lower fabrication cost, an easier multiphase generation, and a better scalability with the CMOS technology scaling.

To be frequency-tunable, a frequency adjustment mechanism should be introduced to evolve an oscillator to a VCO. Both LC and ring oscillators may have two

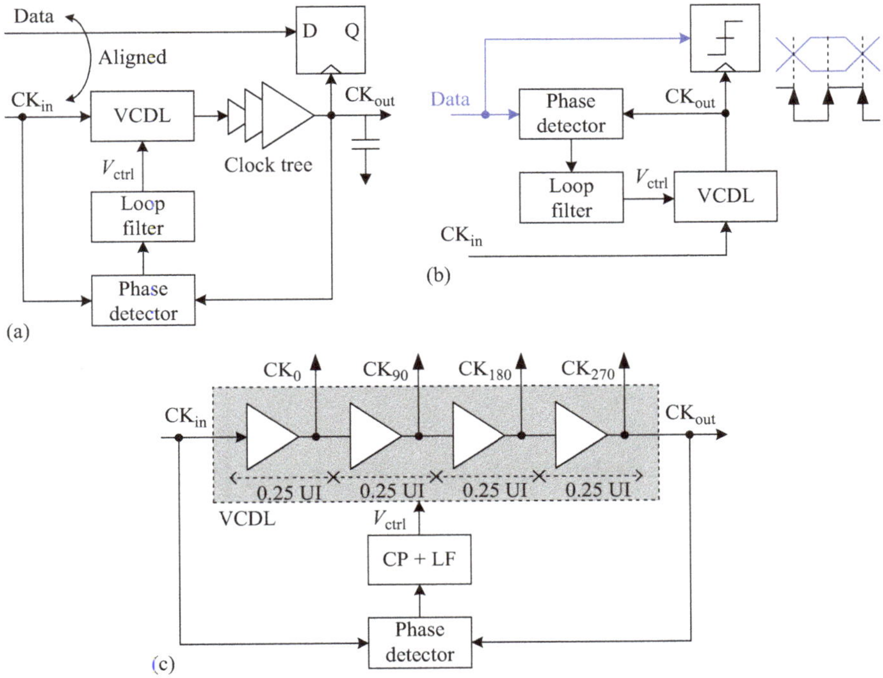

Figure 5.4 DLL used as (a) zero delay buffer, (b) clock and data recovery circuit, and (c) multiphase clock generator

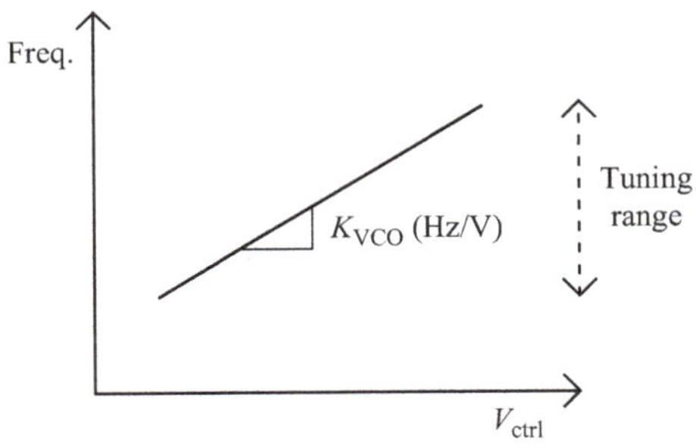

Figure 5.5 Definition of tuning range and VCO gain (K_{VCO})

tuning mechanisms, respectively, as the frequency of an LC oscillator is determined by the product of L and C, and that of a ring oscillator is determined by the CV/I time delay of a delay element. For the LC oscillator, although there have been some examples utilizing inductive tuning [9], generally capacitive tuning is used. It is because an MOS varactor, which provides a high capacitance density and a

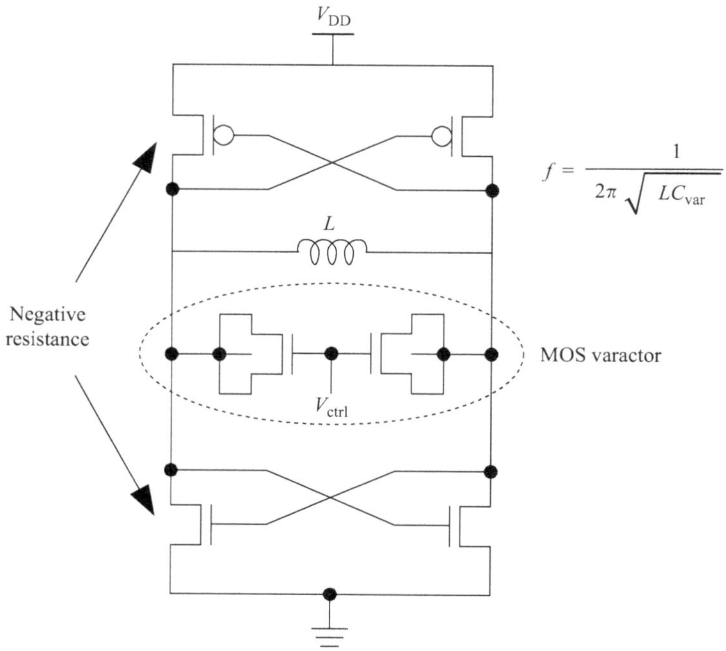

Figure 5.6 Schematic diagram of LC oscillator

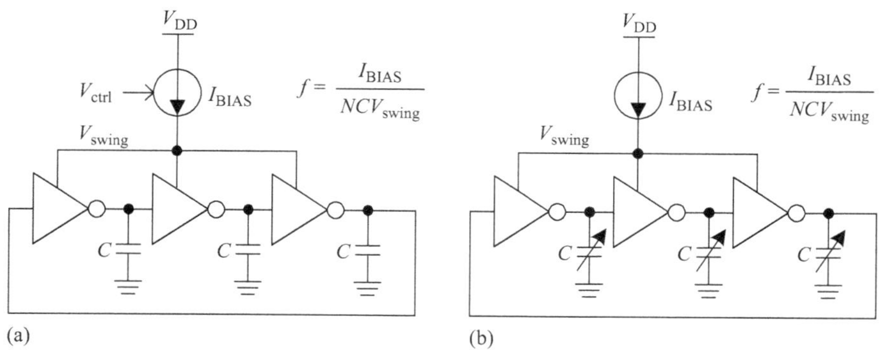

Figure 5.7 Schematic diagram of ring oscillator: (a) current-control scheme and (b) capacitance-control scheme

reasonable tuning range is available in a CMOS technology [10]. The MOS varactor is very similar to a metal-oxide-semiconductor field-effect transistor (MOSFET) structure, so it can also take advantage of the CMOS technology scaling. The gate leakage current and parasitic resistance at source and drain terminals degrade the quality factor of the MOS varactor [11,12]. However, the quality factor of MOS varactor is generally higher than that of an on-chip inductor, thus it does not make a significant difference on the LC tank. The K_{VCO} of the LC oscillator is calculated as

$$K_{VCO} = \frac{df}{dV_{ctrl}} = -\frac{1}{4\pi C_{var}\sqrt{LC_{var}}} \cdot \frac{dC_{var}}{dV_{ctrl}} \tag{5.1}$$

On the other hand, ring oscillator can take either of the current tuning and the capacitive tuning, whose simplified examples are shown in Figure 5.7. Based on the simplified ring oscillator model in Figure 3.12 (the first-order approximation), the bias current I_{BIAS} equals the current flowing to the node capacitor during the transition of each inverter. Therefore, the delay of each inverter can be written as

$$\tau_d = \frac{CV_{swing}}{2I_{BIAS}} \tag{5.2}$$

where V_{swing} is the output voltage swing of the oscillator. As a result, the oscillation frequency becomes:

$$f = \frac{1}{2N\tau_d} = \frac{I_{BIAS}}{NCV_{swing}} \tag{5.3}$$

where N is the number of stages. The K_{VCO} can be calculated as

$$K_{VCO} = \frac{1}{NCV_{swing}} \cdot \frac{dI_{BIAS}}{dV_{ctrl}} - \frac{I_{BIAS}}{NCV_{swing}^2} \cdot \frac{dV_{swing}}{dV_{ctrl}} - \frac{I_{BIAS}}{NC^2V_{swing}} \cdot \frac{dC}{dV_{ctrl}} \tag{5.4}$$

Note that V_{swing} is set to the bias point where the current of PMOS equals I_{BIAS}, so that it is a function of V_{ctrl} and the size of the inverters. As a result, it is not a constant if the current tuning is adopted. V_{swing} is obtained by solving:

$$I_{BIAS} = \beta\frac{W}{L}\left(V_{swing} - V_{TH}\right)^{\alpha} \tag{5.5}$$

where α ($1 \leq \alpha \leq 2$) and β are the technical parameters. Equation (5.5) leads to

$$V_{swing} = V_{TH} + \frac{L}{\beta W}(I_{BIAS})^{\frac{1}{\alpha}} \tag{5.6}$$

where we can find there is a nonlinear dependency between the V_{swing} and I_{BIAS}. From (5.4) and (5.6), we can also find that the K_{VCO} is not a constant even though I_{BIAS} is linearly proportional to the V_{ctrl}. Typically, the K_{VCO} becomes smaller at a higher VCO frequency. In addition, since the I_{BIAS} is a function of the V_{ctrl}, the V_{swing} is not a constant across the tuning range. As a result, generally, a level converter buffer is required to make the oscillator signal to be compatible with the rest of the circuits in a chip. On the other hand, the capacitance-tuned ring oscillator flows a fixed current across the tuning range so that the V_{swing} is a constant. Moreover, the K_{VCO} becomes a constant if the varactor capacitance is linearly

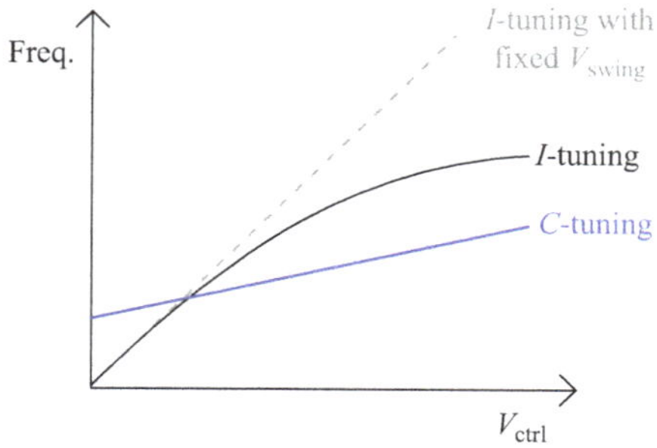

Figure 5.8 Comparison of f–V curves of the current tuning and the capacitance tuning

proportional to the V_{ctrl}. The drawbacks of the capacitance tuning are the limited tuning range by the varactor range and the lower frequency because of the intrinsic varactor capacitance, whereas the current tuning has much wider tuning range and higher frequency. The comparison of the *f–V* curves of the current tuning and the capacitance tuning is illustrated in Figure 5.8, where $\frac{dI_{BIAS}}{dV_{ctrl}}$ and $\frac{dC}{dV_{ctrl}}$ are assumed to be a constant for the current tuning and the capacitance tuning, respectively. While having a wider tuning range, the frequency of the current tuning is saturated at a high V_{ctrl}, because of the increased V_{swing} and limited current density of the transistors.

The schematic diagram in Figure 5.7 includes the pull-up bias current only for simplicity; however, we may need a pull-down bias current for a better matching of the pull-up and pull-down transitions. Recall the LTV phase noise analysis on the ring oscillator; the symmetric transitions suppress the $1/f$ noise. The pull-down current is set by the current equation of the NMOS device in the inverter expressed in (5.5), so it is a function of the I_{BIAS} and the V_{swing}. In average, it must be the same with the I_{BIAS}; however, its dependency on the V_{swing} makes the shape of the current pulse different. It leads to asymmetric pull-up and pull-down transitions. Moreover, the intrinsic differences between PMOS and NMOS (i.e., V_{TH}) also introduces another factor that results in the asymmetricity. Simultaneous use of PMOS and NMOS bias, which is shown in Figure 5.9, can provide a better matching which suppresses $1/f$ noise. However, the bias circuit is another noise source so that its contribution to the phase noise should be carefully considered. Note that 1:N current mirror also amplifies the current noise by a factor of N.

5.2.2 Phase detector

A phase detector is characterized by a phase detector gain curve which shows the average output voltage with respect to the input phase difference. A primitive example of phase detector is an exclusive-OR (XOR) gate shown in Figure 5.10. Since the XOR gate examines if the two inputs have the same binary value or not, it detects how far the positive/negative edges of the two clock inputs are apart from

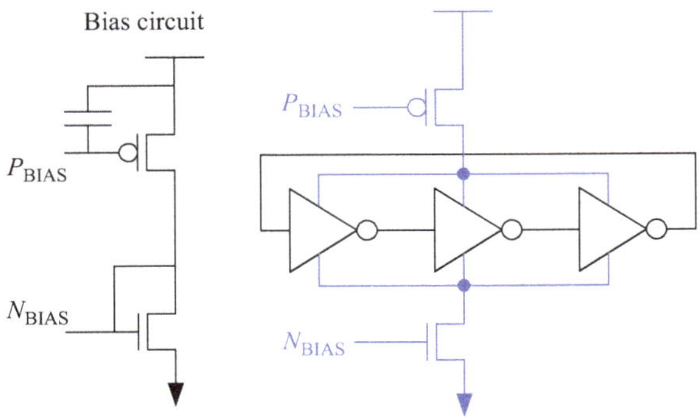

Figure 5.9 Ring-VCO with matched pull-up and pull-down current

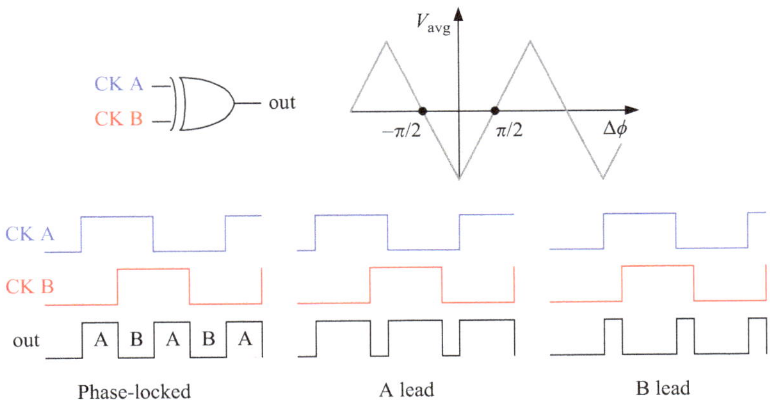

Figure 5.10 XOR phase detector

each other. Therefore, the output pulse-width is linearly proportional to the input phase difference. Such phase detectors which have this linear property are categorized to a linear phase detector. Note that the XOR phase detector output is triggered by the negative edges of two clock inputs as well as the positive edges because it basically detects the input difference according to the input voltage level. Thus, the XOR phase detector output is balanced when the phase difference equals $\pi/2$, instead of zero, so there is a phase offset. Therefore, the XOR phase detector may not be appropriate for some application which requires a precise phase alignment, such as the zero-delay buffer.

The XOR phase detector has another issue of a clock duty cycle because it is level sensitive. As shown in Figure 5.11, if one of the input clocks has a distorted duty cycle of 0.25, the gain curve shifts by $\pi/4$. It is because the phase differences of positive edges and that of negative edges are averaged in the XOR phase detector. Moreover, the XOR phase detector loses the linearity at regions where the

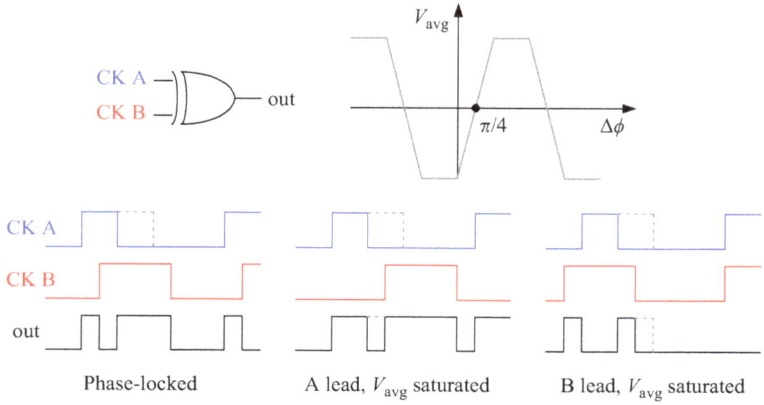

Figure 5.11 XOR phase detector with duty-cycle-distorted input

phase difference is larger than a certain amount because the output pulse-width is constrained by the narrow pulse-width of the clock A. This duty-cycle sensitivity also limits adopting the XOR phase detector in some applications. For example, the output of a frequency divider in the clock multiplier application does not always have the duty cycle of 0.5 depending on the division ratio.

To address the issues of the XOR phase detector related to the reactivity to the input level, a JK latch can be used as an edge-sensitive phase detector. The JK latch is triggered only by the positive edges of the input clock, that is the positive edges at J and K inputs trigger the output to high and low, respectively. As a result, the phase detector output is not sensitive to the voltage level of the input clocks, in other words, the duty cycle, and is affected only by the phase difference between the positive edges. The output is balanced when the positive edge of the clock B is aligned to the middle of the adjacent positive edges of the clock A, that is, at the phase difference of π, so that the JK phase detector is not able to provide zero phase offset as well (Figure 5.12).

An example of the phase detector, called a dynamic phase detector, which has zero phase offset is found in Figure 5.13 [13]. When both CK A and CK B are low, the node D1 and node D2 are charged to supply voltage (V_{DD}), then the PMOS P1 and P2 are turned off. Since NMOS N1 and N2 are turned off by the CK B and the CK A, respectively, the DN and UP keep the previous status. If one of the clocks, for example, the CK A, is triggered to high, D1 is discharged to GND whereas D2 becomes floated but the nodal capacitance dynamically stores V_{DD}. Subsequently, P1 is turned on while P2 is still turned off, and therefore only DN is switched to low. At the same time, N2 and N4 are turned on so that they form a discharging path, and subsequently UP is switched to high. In the same manner, UP/DN will be respectively switched to low/high if CK B is triggered. When both CK A and CK B become high, D1 and D2 are discharged to GND, and subsequently both UP and DN are reset to low. From this state, switching one of the clocks to low does not

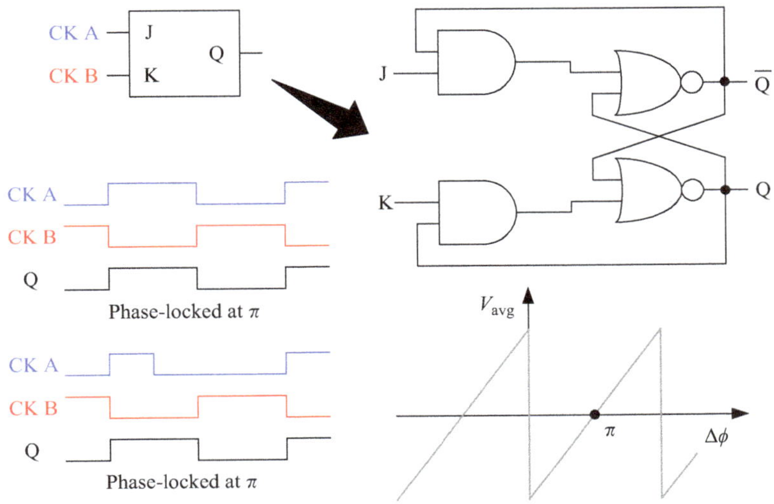

Figure 5.12 JK latch as an edge-sensitive phase detector

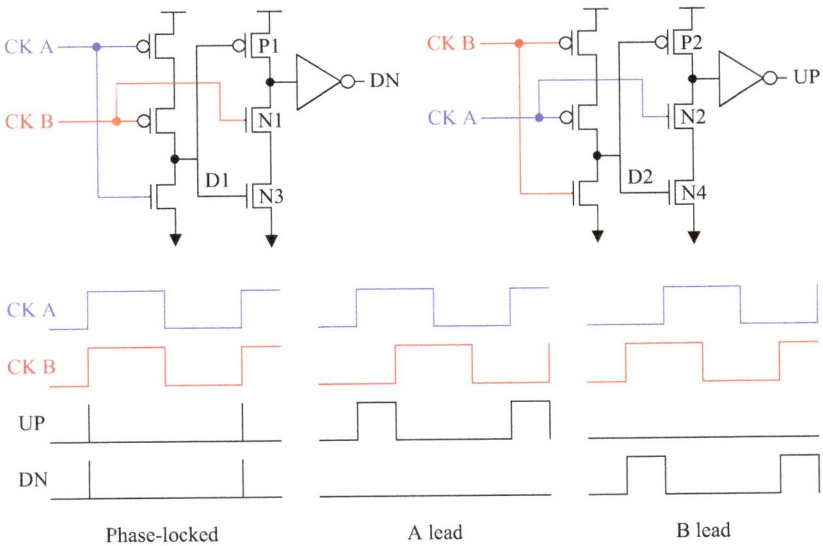

Figure 5.13 Dynamic phase detector

affect the output, since D1 and D2 keep GND, one is still tied through NMOS but the other is floated while keeping the previous state. From the observation, the dynamic phase detector does not react to the negative edge and provides the zero-phase-offset detection, because it detects how far the positive edges are apart from each other. The dynamic phase detector cannot be used at very low frequency since leakage current may flip the dynamically stored voltages.

Figure 5.14 shows a different kind of phase detector which has the zero-phase offset and is also insensitive to the negative edge. Implemented with a single D-flipflop, this phase detector samples one clock by the other clock. The output of the phase detector shows whether the sampled clock is high or low at the instant of the sampling, that is, the positive edge of the sampling clock. As a result, it can distinguish if the sampling clock leads or not, however, is not able to detect how much it does. This kind of binary quantized phase detector is referred to as a bang-bang phase detector. Despite the shortcomings from the nonlinear nature, the bang-bang phase detectors are widely used because of its high-speed capability; it can work at the highest speed that the D-flipflop works in each CMOS technology.

Note that the phase detectors described above do not operate if NRZ data sequence is applied. For the CDR application, the phase detector should be able to detect the phase difference between the VCO clock and the NRZ data [14]. In order for supporting such operation, different types of phase detectors should be introduced. Figure 5.15 shows two phase detectors that compare the phase difference between an NRZ data and a clock. The first phase detector, which was first proposed by Charles R. Hogge in [15], exhibits a linear gain. Assuming the full-rate clock whose frequency equals to the twice of the Nyquist frequency of the incoming NRZ sequence (i.e., the Nyquist frequency of 10 Gb/s NRZ is 5 GHz), the NRZ is sampled by every clock cycle (X) and then is delayed by a half-clock period by the second D-flipflop (DFF). As a result, every time the NRZ has a transition, the Hogge phase detector generates the UP pulse whose width equals to a half-clock period. On the other hand, the phase delay from Data to X is given by the phase difference from the transition of Data to the positive edge of CLK, since X is retimed by CLK. Therefore, the pulse-width of DN is the same as the phase

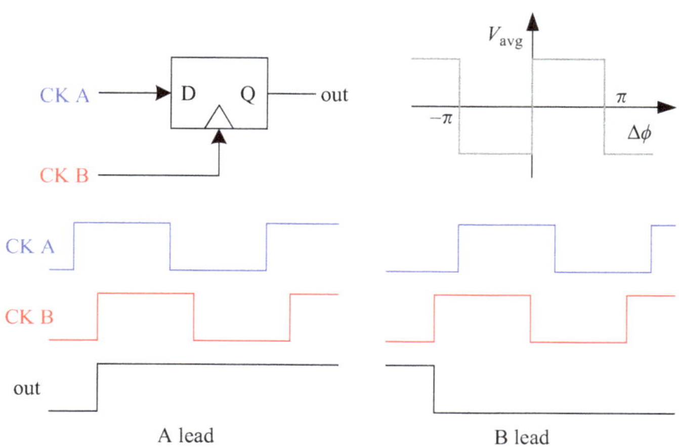

Figure 5.14 Bang-bang phase detector

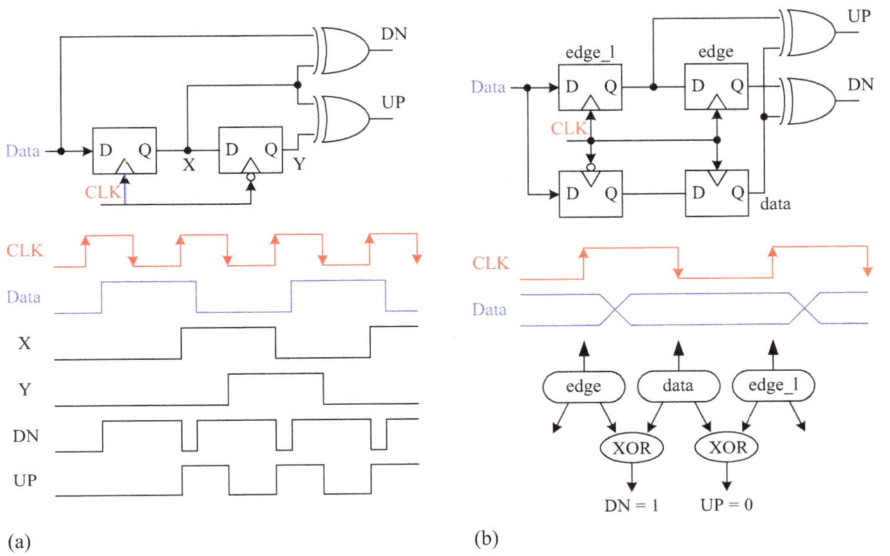

Figure 5.15 Phase detector with NRZ data: (a) Hogge phase detector and (b) Alexander phase detector

difference between Data and the positive edge of CLK so that net output ($UP_{avg} - DN_{avg}$) is proportional to how large the phase difference is. We can easily find that the pulse-width of DN and that of UP are balanced only if the CLK positive edge is aligned to the center of the data unit-interval (UI), where the maximum sampling margin is achieved.

One of the drawbacks of the Hogge phase detector comes from the fact that the NRZ data are directly compared to the sampled signal through the XOR gate, which is responsible for the DN. That is, the data are required to be "processable," which means the common-mode and voltage swing are compatible with the DFF output and the signal quality is clean enough. However, such conditions cannot be met as the data rate goes higher, which implies that the Hogge phase detector is not appropriate for high-speed applications. The Alexander phase detector shown in Figure 5.15(b) fits better to high-speed, at the cost of losing the linearity [16,17]. Because the Alexander phase detector uses the sampled value rather than comparing the incoming data directly, the exact phase information of the data cannot be used. Instead, the data are sampled twice per UI utilizing both positive and negative edges of the full-rate clock. Typically, the samples from the positive edge and from the negative edge are referred to as the edge sample and the data sample, respectively. The DFFs in the second column are used to store the previous value of the edge sample and to retime the data sample to the positive edge of the clock for the XOR logic. If the edge sample and the data sample have a different value, that is, when the output of the XOR gate goes high, we can find that the data transition has occurred somewhere between the time when the edge and data samples were gathered. Therefore, by comparing the data sample to the two adjacent edge samples, we can detect whether the edge sample is collected before the data transition or not. Figure 5.15(b) shows the example of waveforms and operation when the edge sample is collected before the data transition.

For a PLL where the VCO frequency is adjusted to correct the phase, the frequency information should also be gathered and corrected by the loop. However, the phase detectors described above are not able to detect the frequency difference. For example, assume that there is a frequency offset between CK A and CK B of the XOR phase detector as shown in Figure 5.10. Because frequency offset is equivalent to phase drift, the phase difference detected at each clock cycle will drift. Therefore, the phase detector output sweeps the gain curve and is accumulated at the loop filter. Note that the sweeping and accumulation corresponds to the integration. The integration of the gain curve must be zero, which implies that the average output of the phase detector equals zero unless there is a frequency offset. The PLL can lock the loop only when the frequency offset is small enough, such that it can be corrected within a short period of time before fully averaged. If the offset is very small, the phase step of the sweeping becomes very slow so that the integration of the phase detector output stays nonzero for a while. If this time is long enough that the PLL can correct the frequency offset, the PLL acquires the phase/frequency lock. The maximum frequency offset that a PLL can tolerate is called an acquisition range. The acquisition range is a function of the PLL loop gain; however, it is typically not very wide. Moreover, the phase detectors cannot distinguish harmonic frequencies. For example, let us assume that the frequency of one clock is twice higher than the other clock as shown in Figure 5.16. As long as the edges of the clocks are aligned, the average of the phase detector output becomes zero. The PLL thinks the input clock and the output clock is perfectly matched, even though the output clock frequency is twice higher (or lower) than the input frequency. This specific case is known as a harmonic locking.

To avoid the limited acquisition range and the harmonic locking, a phase/frequency detector (PFD, Figure 5.17) can be used alternately. The PFD is composed of two resettable D-flipflops (reset to low) and an OR gate. Because both D-flipflops take V_{DD} to the input, they simply trigger the output to high when a positive edge of the sampling clock rises. In addition, once both outputs go to high, the feedback path returns them to low after the feedback path delay. Therefore, if the CK A triggers first, the UP becomes and stays high until DN is triggered by the positive edge of CK B regardless of how many edges of the CK A have been triggered until then. Therefore, the PFD can distinguish which clock triggers earlier and also detect how far they are, so it exhibits a linear gain curve. Moreover, it is important to notice that the PFD remembers the edges which it has been storing and comparing. That is, the PFD remembers the order of edges and compares the edges in the same order from the first comparison, whereas the phase detectors compare the nearest edges regardless of the order.

We can observe that very narrow pulse will be asserted when the phase difference is small enough. However, due to the transient nonideal effect such as a finite rising/falling time, too narrow pulse cannot propagate through a CMOS

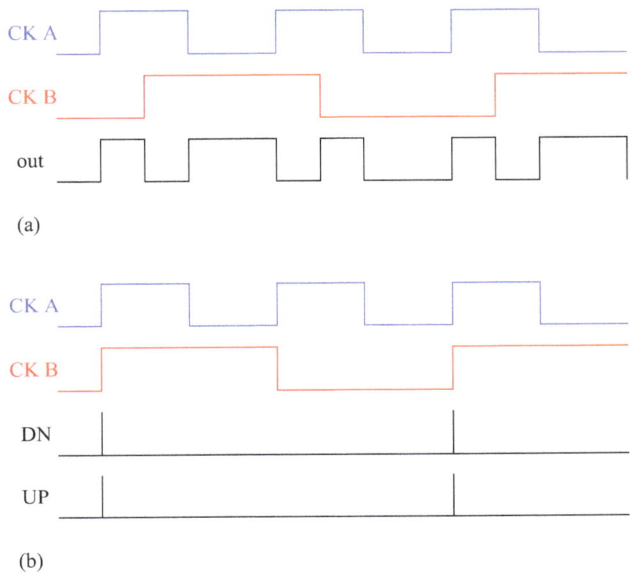

Figure 5.16 *An example of phase-detection failure due to frequency offset: harmonic locking (a) XOR PD: Balanced output with frequency offset (b) Dynamic PD: Balanced output with frequency offset*

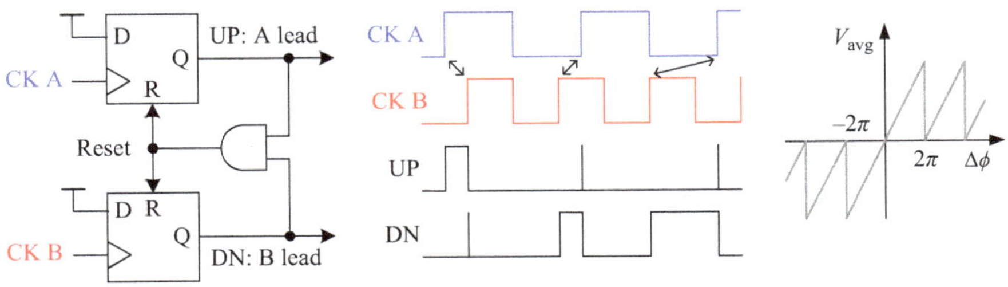

Figure 5.17 *Phase/frequency detector*

circuit. That is, if the pulse-width decreases as the phase difference decreases, the PFD does not produce any output pulse for a small amount of the phase difference. It leads to a flat region in the gain curve, which is named "dead zone". Within the dead zone, the PFD does not tell how large the phase difference is, so that the PLL does not lock at the exact zero phase difference but also it is not able to detect phase drift until the phase difference becomes larger than the dead zone, which results in jitter. A practical remedy on the dead zone issue is to introduce additional delay to the feedback path, as shown in Figure 5.18(a). With the additional delay τ, UP/DN signals have enough time to retain the pulse before the feedback resets them. The drawback of the additional delay is the reduced frequency acquisition strength.

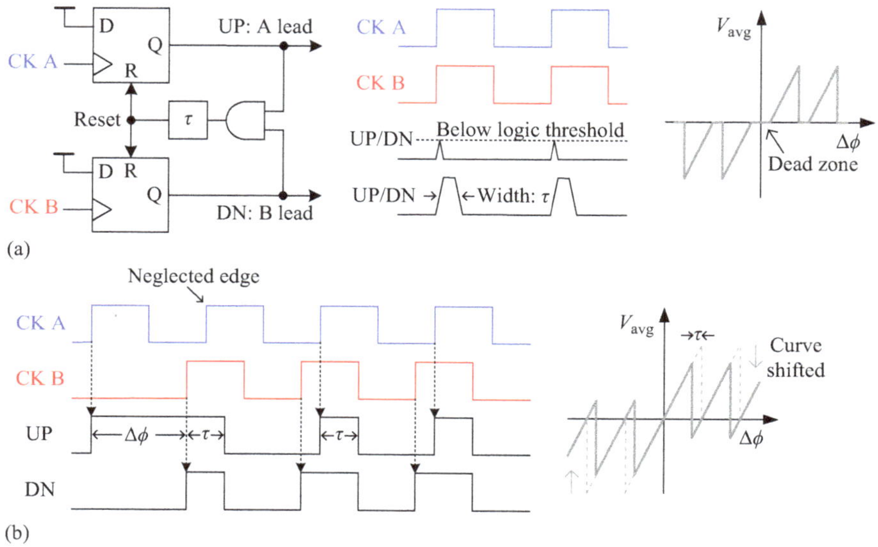

Figure 5.18 *(a) Dead zone issue of PFD and (b) gain curve shift by the reset path delay*

That is, when the phase difference Δø is large, for example around 2π, the sum of Δø and τ may exceed 2π. Then the second positive edge of the leading clock (CK A) arises before UP/DN signals are reset, and therefore it cannot be counted. In the next cycle, the PFD compares the second edge of CK B with the third edge of CK A, which leads that the detected phase difference is shifted by -2π from the actual value. It introduces the negative regions in the PFD gain curve, which results in degraded frequency acquisition capability [18]. For an extreme case that the clock frequency becomes too fast such that τ exceeds π, the PFD is no more able to detect the frequency correctly since the negative region becomes larger than the positive region. Moreover, the pulse-width needs to be minimized as far as the dead-zone is avoided because the noise of the charge-pump is injected only when the charge-pump is turned on. This aspect will be discussed detailed in Chapter 8 and in Appendix A, where we will derive the phase noise contribution of the charge pump and the PFD.

We found that the PFD is amazingly useful for PLL implementation. On the contrary, we should be careful of using the PFD for the DLL. The frequency acquisition capability, which is based on the fact that the PFD remembers the order of edges, may not be useful for the DLL, where the output frequency is always the same as the input frequency. Figure 5.19 explains the start-up issue that arises when a PFD is used in a DLL. Because of the nature of the "delay," the CK B is always a delayed version of CK A. So for a correct comparison, the first edge of the CK A needs to be neglected and the second edge should be compared to the first edge of the CK B. However, if there is no additional circuit to let the PFD neglect the first edge of CK A, the PFD will compare the first edge of the CK A to that of the CK B and assert UP pulse regardless of the actual phase difference.

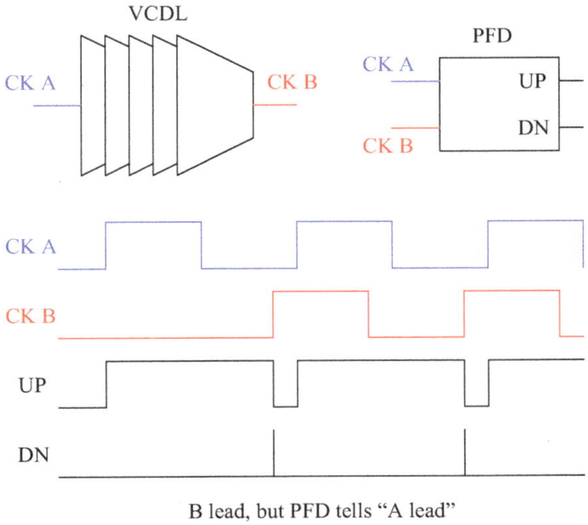

Figure 5.19 PFD start-up issue in DLL

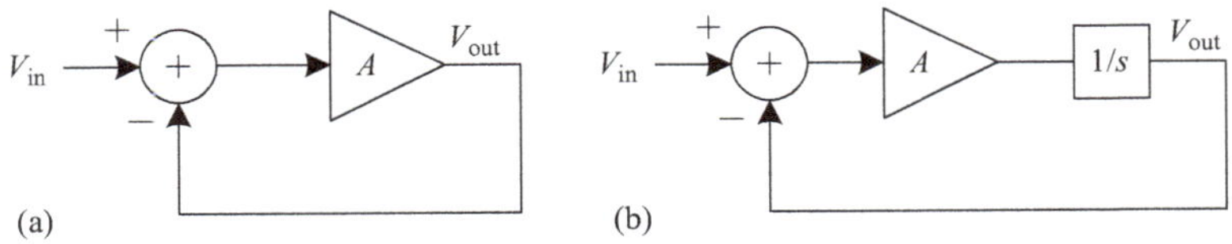

Figure 5.20 Feedback system (a) without integration and (b) with integration

5.2.3 Charge pump and loop filter

The loop filter determines how to control the VCO/VCDL from the phase error information gathered from the phase detector. The design of loop filter is very important because it is designed at the final design step to set the loop characteristics such as the transfer function and the loop stability from the given circuit parameters of the phase detector, the VCO/VCDL, and the frequency divider. The charge pump, which proportionally converts the phase error from the phase detector into the amount charges injected to the loop filter, has been widely used to implement the loop filter for PLL/DLL. Figure 5.20 shows a brief background on the loop filter in a feedback system, where we can find why such charge-pumps have been widely used to implement the loop filter. For a negative feedback system whose open-loop gain is A, the closed-loop gain is written as

$$\frac{V_{out}}{V_{in}} = \frac{A}{1+A}. \tag{5.7}$$

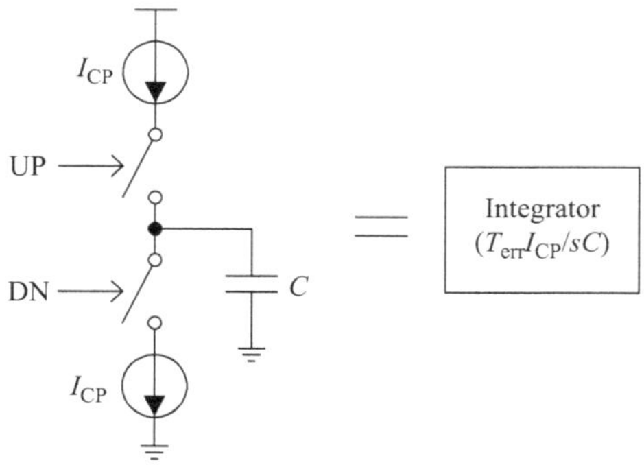

Figure 5.21 Integrator implementation with a charge-pump and a capacitor

Equation (5.7) implies that the input and the output have a static offset of $1/(1+A)$. Placing an integrator in the loop makes the DC loop gain be infinite, and therefore the static offset is eliminated.

Now let us see how an ideal integrator is implemented in integrated PLL/DLLs. The phase detector or the PFD output contains the phase error information in the duration of the pulse (T_{err}). Implemented with a pair of a current source and a switch as shown in Figure 5.21, the charge pump flows a fixed current (I_{CP}) for the pulse duration so that it converts T_{err} to charge ($I_{CP}T_{err}$). Because a capacitor is a storage element for the charge, the charge-pump followed by a capacitor is served as an ideal integrator.

One of the most important design issues of the charge pump is the mismatch between the pull-up current and the pull-down current, which causes a static phase offset and a spurious tone at the frequency offset of the reference frequency. Detailed analysis on these issues will be discussed in Chapter 6. Figure 5.22(a) shows a simple implementation of the charge pump with a bias circuit. Here, vn1 is assumed to be provided from an on-chip or external bias generator. To mimic the charge-pump current, the bias circuit is implemented by the replicas of the transistors composing the charge pump. Because there is only one current path, the vp1 is automatically set to match the PMOS current to be the same as the NMOS current (I_{BIAS}) by the diode-connected transistor. Because the charge pump takes vp1 as well as vn1, the pull-up current (I_P) and the pull-down current (I_N) seem to be the same. However, due to the finite output resistance of the charge pump, the I_N increases as the V_{ctrl} increases but I_P decreases. I_N and I_P do not equal to I_{BIAS}, because the diode-connected bias vp1 is not the same as V_{ctrl}. A feedback biasing shown in Figure 5.22(b) can be used to keep the drain voltage of the bias circuit to V_{ctrl}. As long as the amplifier gain is large enough, the drain voltage is almost the same as V_{ctrl}, the bias circuit has the same nodal voltages as well as the transistors. So, I_N and I_P equal to I_{BIAS} regardless of V_{ctrl}. However, the I_{BIAS} is not a constant since it still increases as the V_{ctrl} increases. It may result in uncertainties to the loop dynamics that will be discussed later. Adding another compensation feedback to the bias circuit and one more pump branch as shown in Figure 5.22(c) can mitigate

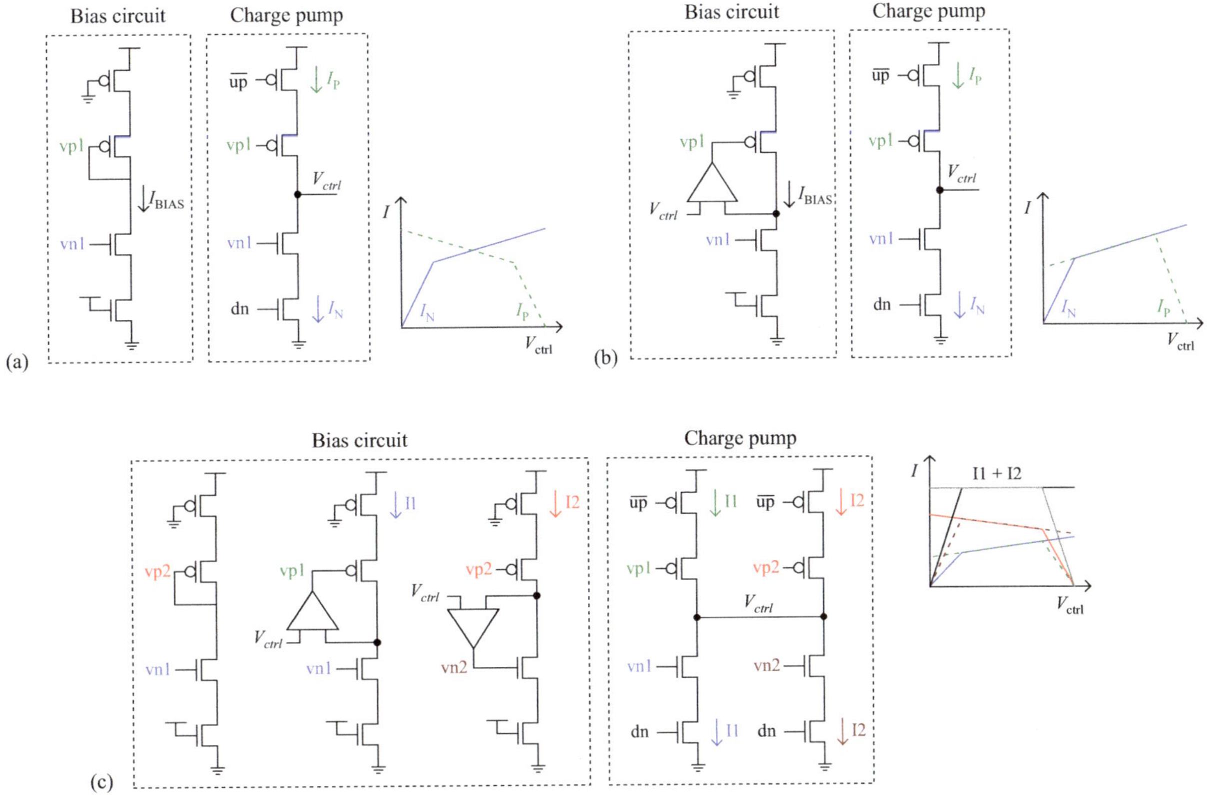

Figure 5.22 Circuit diagram of charge pump with (a) simple bias circuit, (b) single compensation bias circuit, and (c) double compensation bias circuit

the V_{ctrl} dependency. Using the same bias circuit in Figure 5.22(a), the PMOS bias voltage vp2 is self-generated by the diode-connected PMOS from the vn1. The same replica-feedback bias circuit from Figure 5.22(b) produces vp1, which tracks vn1 with respect to the V_{ctrl}. The second feedback circuit does the same thing what the first feedback circuit is doing for replicating vn1, that is, it generates another pair for vp2. Note that the current flowing by the pair of [vn1, vp1] increases as V_{ctrl} increases, whereas the current from the [vn2, vp2] pair decreases. As a result, when we have two current branches as shown in Figure 5.22(c), the net V_{ctrl} dependency of the pull-up and pull-down currents are canceled out each other.

The loop filter design for a PLL is generally more complicated than that for a DLL. It is because the PLL adjusts the frequency of the VCO to correct the phase error. Therefore, the VCO introduces a pole at the zero frequency in the loop phase transfer function since the phase is the integration of frequency. To get an insight into the PLL loop filter design, let us see four possible loop filter implementations shown in Figure 5.23. A capacitor-only loop filter preceded by the charge-pump is an integrator, so the PLL has two poles at the zero frequency which leads to zero phase margin* if such loop filter is used. On the other hand, the resistor-only configuration does not introduce an additional pole, so the loop is always stable because there is only one pole in the loop by the VCO. However, as we studied the PFD, the phase error should be integrated over time to extract the frequency difference. For example, if the VCO frequency is lower than the reference frequency, the PFD asserts UP at every clock cycle, but the VCO control voltage instantaneously rises but returns to the initial voltage once the UP pulse is reset. That means the PLL tries to track the frequency by updating the phase at every reference cycle while keeping the VCO frequency a constant. As a result, the PLL can only track a very small frequency offset so that the frequency acquisition range is very narrow. The combination of R and C gives a practical solution. As well as an instant IR drop by the resistance, which is finally converted to phase shift

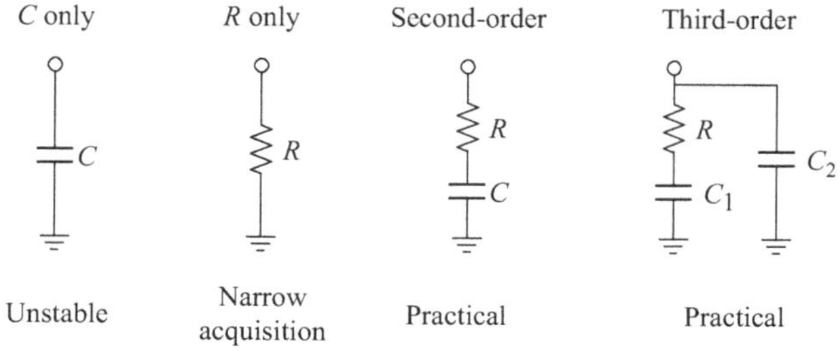

Figure 5.23 PLL loop filter examples

*Phase margin is an indicator of the stability of a negative feedback. It shows how far the negative feedback system is apart from the unstable oscillation, that is the Barkhausen Criterion. The phase margin is defined to be the phase difference between 180° and the phase of the open-loop transfer function at 0-dB gain.

by VCO, the capacitor integrates the charges from the pump to track the frequency error. From the view of pole-zero interpretation, the resistance introduces a zero in addition to the two poles so that the PLL can retain an adequate phase margin. The single resistor-single capacitor loop filter is generally referred to as a second-order loop filter because the loop transfer function becomes second-order when the loop filter is combined with the pole from the VCO. The third-order loop filter configuration includes another parallel capacitor. This additional capacitance smooths the instant IR ripple due to the resistance hence mitigating a spurious tone at the output of the VCO. The detailed analysis of the PLL loop transfer function will be discussed in Chapter 6.

5.2.4 Frequency divider

In a PLL, frequency divider is the circuit which should operate at the highest speed along with the VCO. Specifically, the first stage of the divider should work at the highest frequency that a VCO can produce; otherwise, the PLL will fall into a false locking state. The false locking issue will be discussed in Section 5.4. The easiest way to implement a frequency divider is the use of a DFF in negative feedback [19], which achieves frequency division by two. Figure 5.24 shows examples of frequency divider with a DFF. Using a master–slave DFF shown in Figure 5.24(a) might be the most common way. Assume the output node (*out*) is low at the initial state and the *clk* is low. Also note that a latch is only transparent when the clock is high, if not it holds the previous value. Once the *clk* goes to high, the first latch becomes transparent while the second latch does not, so "low" state propagates up to node *A*. At the next falling edge of the *clk*, the "low" stored in the *A* goes through the second latch and is finally inverted by the output inverter. At the next rising edge, the first latch is transparent so that the stored value on the node *A* is flipped. We can find that it took one clock period to toggle the latch output, which means that the frequency is divided by a factor of two. The operating speed of the divider is generally limited by the clk-to-q delay and the setup time of the latch and the clk-clkb mismatch is introduced by the inverter. Using current-mode logic (CML) latch shown in Figure 5.25 can boost up the speed by reducing the signal amplitude [20]. However, compared with CMOS latches, the CML latch consumes more power because of the static current dissipation. In addition, it also requires a bias circuit and a fully differential signal which increase hardware overhead. These aspects let the CML divider to be adopted in high-end applications. Dynamic implementations of the latch and the DFF provide alternative solutions. In Figure 5.24(b), a true-single-phase clock (TSPC) divider is shown. The TSPC flipflop stores the voltage on the parasitic capacitance, unlike a static latch where the voltage is stored in the cross-coupled inverter pair. There are only three stages in the divider feedback loop, as a result, the TSPC flipflop is faster than the master–slave flipflop. Moreover, its power consumption is generally lower than those of CMOS static divider and CML divider, since it does not dissipate a static power and has a simple structure. It is also free from clk-clkb mismatch because it uses only one phase clock. However, there are a few downsides of the TSPC approach; first, a full-swing input is needed. Second, it does not work at very low speed because

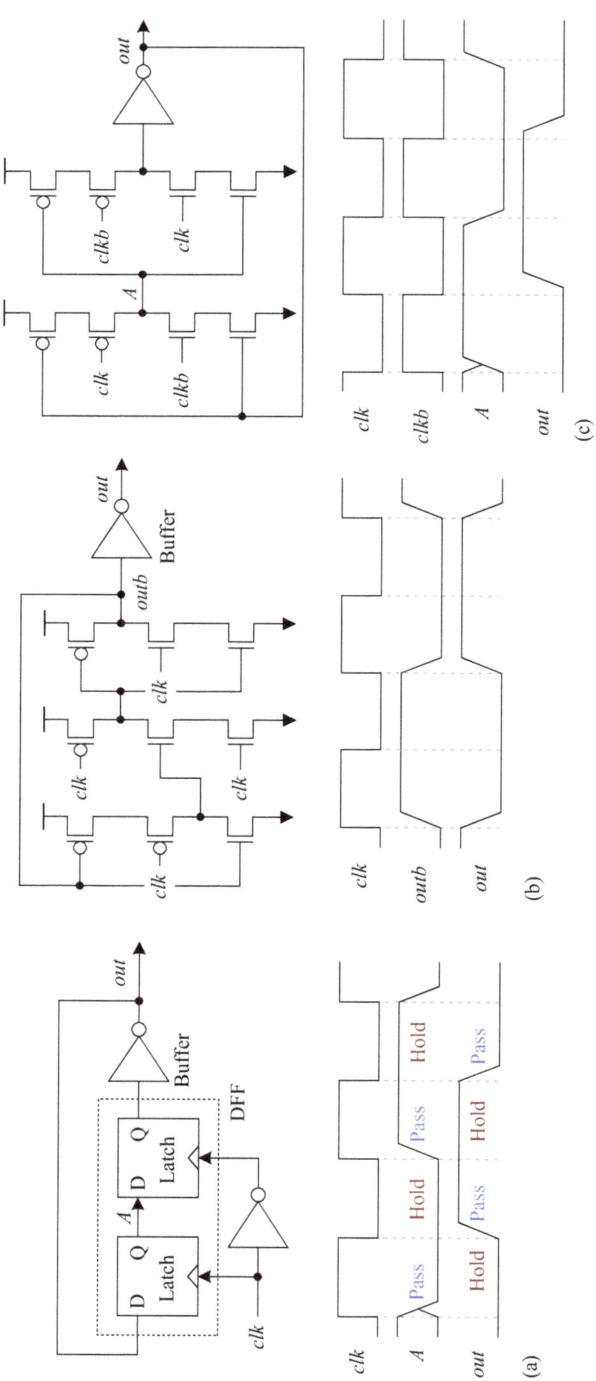

Figure 5.24 Divided by two implementations with D flip-flops: (a) with master–slave D flip-flop and (b) with true-single-phase clock (TSPC) flip-flop

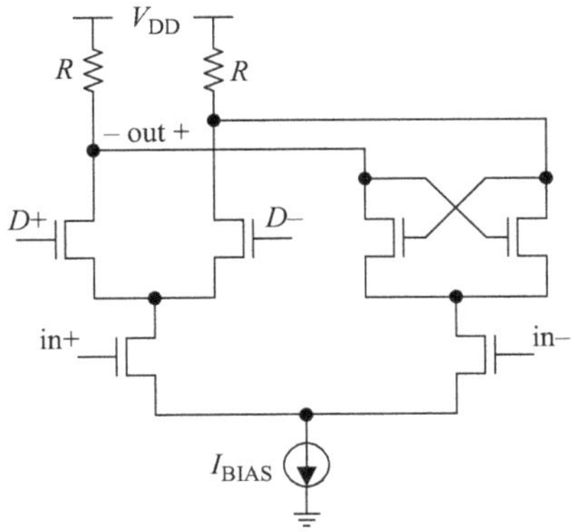

Figure 5.25 Current-mode logic (CML) latch

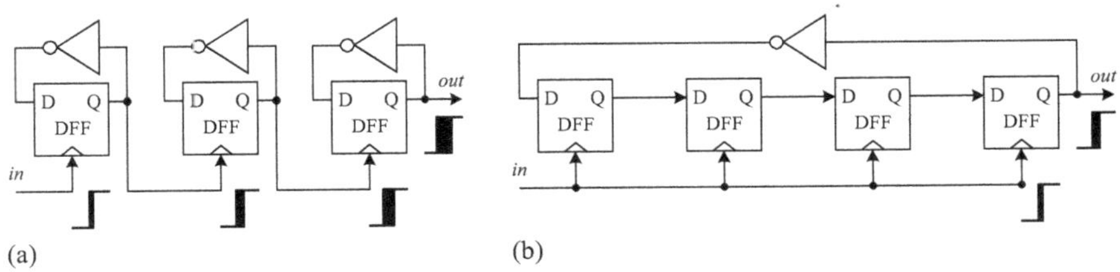

Figure 5.26 Cascaded divider for high divide factor: (a) asynchronous divider and (b) synchronous divider

leakage current disrupts the stored voltage on the capacitor. Third, the stacked PMOS degrades the speed, but in the latest technologies, PMOS is no longer slower than NMOS, so it may not be a problem. Fourth, there is a large load at the feedback node (*outb*) while the stacked NMOS in the third TSPC stage should drive that. Note that the third stage should drive both the first stage and the output buffer. The fourth issue can be addressed by using a dynamic implementation based on a tri-state-inverter latch shown in Figure 5.24(c) [20]. There are three stages in the loop which are the same as the TSPC divider; however, the stacked device is responsible for driving the output buffer only and the feedback node is driven by nonstacked buffer. As a result, the tri-state-inverter-based divider achieves a higher speed than the TSPC divider. However, the differential clock (*clk* and *clkb*) should be aligned well to avoid transparency during clock transitions, so it increases the design complexity [21].

Higher divide factors can be created by cascading the unit dividers as shown in Figure 5.25(a). It is called asynchronous divider because the final output is not synchronized to the input clock. The input clock is sequentially divided into div2, div4, and div8 in the case of divide-by-8. The jitter added by the DFFs is accumulated at the output. Moreover, the clk-to-q delays are also accumulated in the total delay, which degrade the loop latency of a PLL. The counterpart of the asynchronous divider is the synchronous divider shown in Figure 5.26(b). Rather than using a single flipflop as in the divide-by-2 circuit, multiple flip-flops are cascaded as a shift register. We can find that the output of the N-stage synchronous divider is toggled at every $N \cdot T$, so the divide ratio is $2N$. Note that the divide ratio of the asynchronous divider is 2^N. Because the output is retimed by the input clock, the jitter is not accumulated in the synchronous divider. However, the input clock drives all flipflops, which is not the case of asynchronous divider, while the input frequency is the highest in PLLs. Definitely, a large loading on high-frequency clock results in high power consumption.

5.3 Fractional-N PLL

To obtain a desired output frequency with a fine frequency resolution, an integer-N PLL should use a slow reference clock since the frequency resolution is limited to the reference frequency. Dividing the reference frequency shown in Figure 5.2(a) provides frequency multiplication of a fractional ratio, it reduces the effective reference frequency of an integer-N PLL and degrades jitter performance of the PLL. This aspect will be theoretically analyzed in Chapter 6 and also be verified through the survey on the state-of-the-art PLLs in Appendix B. An alternative solution is to use a fractional-N divider. However, as we studied, a frequency divider uses edge information of the input clock. To build a fractional-N divider, we need sub-1-UI spaced edge information which is not available. A practical solution to enable a fractional-N division is to alternate the division factor between two adjacent integer numbers, where the ratio of the durations of each number determines the fractional N. A simplified block diagram of a fractional-N PLL is given in Figure 5.27(a), where FCW, DSM, and MC stand for "frequency control word," "delta-sigma modulator," and "modulus control," respectively. DSM is generally used to convert the fractional FCW to a modulated integer word whose average in time equals to the FCW [22]. For example, with FCW of 4.25, the MC dithers back and forth between 4 and 5 but stays at 4 by three cycles but at 5 by one cycle. Dithering of the division factor is enabled by the use of a dual-modulus divider. Figure 5.27(c) shows an example of $\div 2/3$ prescaler, which is a reconfigurable frequency divider. When the MC goes high, the output of OR gate (B) is always high so the output of the AND gate (C) becomes the same to the *Out*. As a result, it is a divide-by-2 configuration. On the other hand, when the MC is toggled to low, the *Out* stays at high by two cycles but at low by one cycle, which means the configuration is switched to divide-by-3. In Figure 5.27(b), we can find signal waveforms of a fractional-N PLL at a locked state, where an example of divide-by-2.5 is illustrated. Although the average period of the divided clock (*div*) is T_{in}, when we focus on a single period, it dithers between $0.8T_{in}$ and $1.2T_{in}$. The resulting deterministic jitter has $0.4T_{in}$ peak-to-peak. Because the average of the phase error should be nullified at a locked state, the PLL aligns the reference clock to the middle of the dithering. This dithering phase error introduces periodic ripple on the PLL output frequency which degrades the spectral purity. It is referred as to fractional spur. Because of the deterministic nature of the fractional spur, it can be suppressed by using cancellation techniques [23].

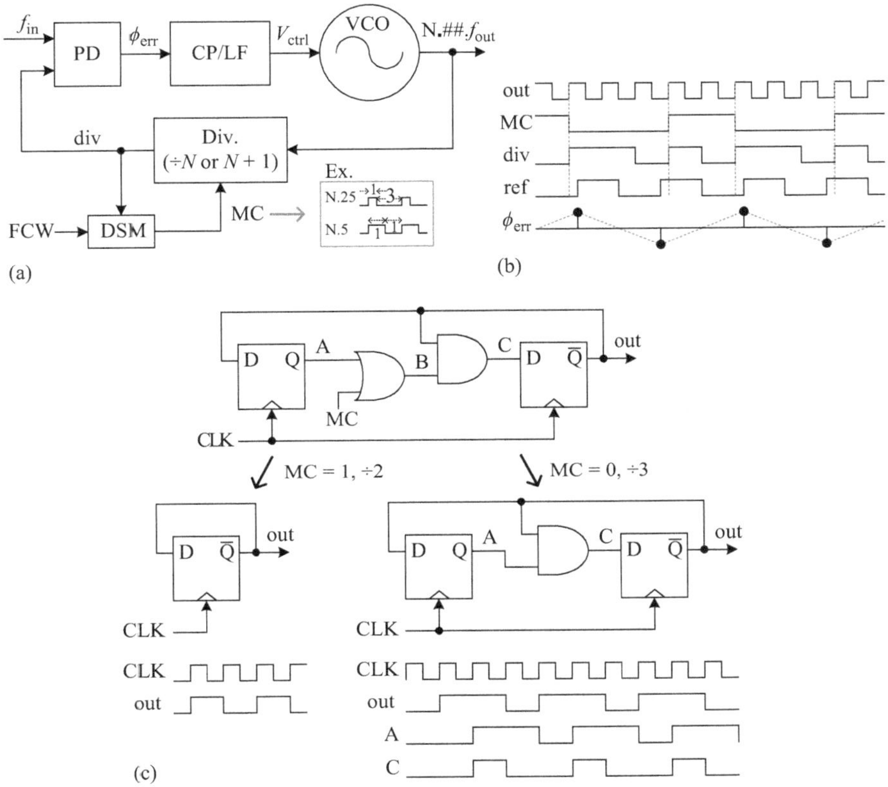

Figure 5.27 (a) Simplified block diagram of fractional-N PLL, (b) waveforms with N = 2.5, and (c) ÷ 2/3 dual-modulus prescaler

5.4 False locking and failure issues in PLL/DLL

Depending on the initial state at start-up, PLL/DLLs may converge to a false lock state. Note that the start-up issue can happen in many other feedback circuits, so sometimes a designer needs to design a start-up circuit. One famous example of the PLL false locking is due to the divider failure. At the start-up, we often don't know where the control voltage of the PLL is. That means, VCO may have very high initial frequency, which is higher than the desired frequency under locked. Assume that the start-up frequency is too high for the divider, that is, the clock pulse-width is narrower than the clk-to-q delay of the latch. Then the divider output does not toggle, so the frequency of the divided clock becomes less than the desired frequency (for an extreme case, 0 Hz). Figure 5.28(b) shows one of the most frequently observed divider failures when a divide-by-2 divider periodically misses the rising edge of the input clock so the division ratio becomes three instead of two. The PFD compares the divided clock with the reference clock and concludes that

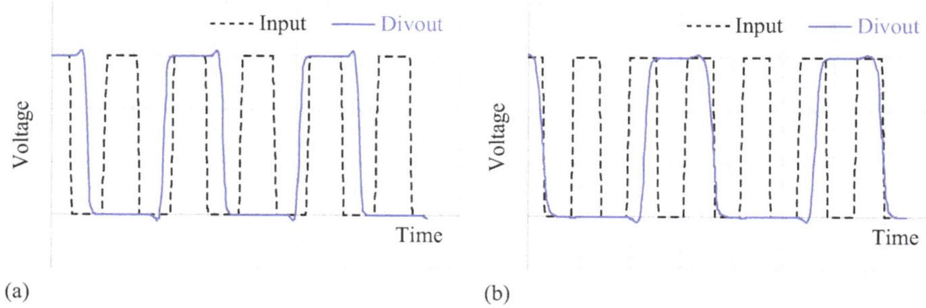

Figure 5.28 Input-output waveforms of divide-by-2 circuit (a) when divider works normally and (b) when divider failure happens due to too fast input frequency

the divided clock is too slow, so the PLL tries to raise the VCO frequency. The VCO frequency will eventually go to the upper limit of the VCO, and there the divider will not go into the correct operation.

Another important convergence failure in PLL happens due to the start-up failure of VCO. Especially, an even-stage ring oscillator, which is frequently used to generate evenly spaced quadrature phases, is vulnerable to such start-up failures. Recall that the Barkhausen criteria are necessary but not sufficient conditions for sustainable oscillations. Even though the ring oscillator satisfies the Barkhausen criteria, depending on the initial condition of the circuit, it can fall into DC equilibrium region and therefore does not oscillate, or enter to false oscillating condition such as common-mode oscillator or complex oscillation [20,24–26]. It is also known that the start-up failure highly depends on the size ratio between the main inverter and the latch inverter. Simply speaking, from the circuit diagram of a four-stage ring oscillator shown in Figure 5.29(a), if we ignore the latches there are eight inverters so it is a latch. Note that when it is latched, the differential nodes (which are supposed to be) have the same binary value. When the cross-coupled latch strength increases, it introduces torque enforcing the nodes to be differential. As a result, the initial condition space that leads to the start-up failure shrinks as shown in Figure 5.29(b). With a sufficiently small failure region with large latch, a small perturbation such as circuit noise or supply ramp is enough to bring the circuit out of the failure region therefore it is free of failure. Since there is a design trade-off that the larger latch reduces the possibility of failures but degrades the speed of the oscillator, careful choice of the latch size is recommended. If the latch size is too large (i.e., $>2.5\times$ of the main inverter) to surpass the strength of the main inverter, the oscillator also fails; however, it is not a practical case.

In a DLL, false lockings are much easier to happen. There are two famous false lockings: harmonic locking and stuck locking. The harmonic locking is attributed to the fact that a DLL adjusts "delay" according to the information gathered from the "phase" detector. Since the phase is periodic, different delays may correspond to the same phase. Figure 5.29 shows an example of DLL harmonic locking. The

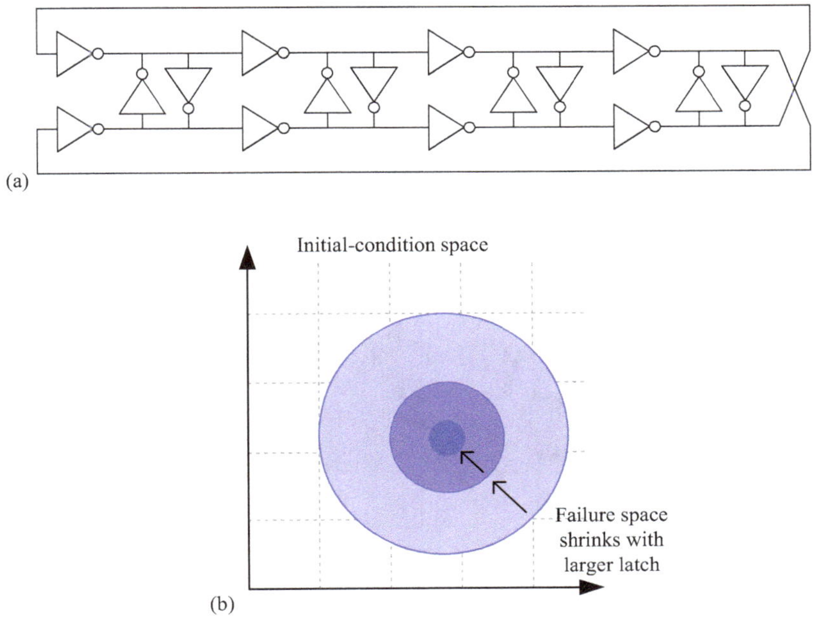

Figure 5.29 (a) Circuit diagram of four-stage differential ring oscillator and (b) conceptual illustration of start-up failure region in initial-condition space

phase detector gain curve is drawn in Figure 5.29(a), where the available range covered by the VCDL is marked in bold straight line. If the range is wide enough, there would be multiple potential locking points. For example, assume that we want to build a multiphase clock generator with a DLL like Figure 5.29(b). Using four-stage delay lines and making the CK_{out} be aligned with CK_{in} at 1-UI delay, which is the left locking point in Figure 5.29(a), four phases (0°, 90°, 180°, and 270°) can be generated. However, if the DLL starts at the initial delay longer than 1.5 UI, the phase detector drives the DLL toward the right locking point of 2-UI delay. Because the adjacent stages are 0.5-UI apart from each other, the harmonic-locked DLL produces only two phases. Limiting the VCDL range can prevent the harmonic locking; however, generally it is not easy to limit the range due to the PVT variations. It becomes even more difficult if the DLL should cover a wide frequency range. There have been several practical and robust techniques presented in literatures, such as delay detector and harmonic lock detector; the readers can consult with [13,27–29] if interested in.

Figure 5.30 shows an example of detecting the harmonic locking, which utilizes multiple phases produced from each of the delay stages [28]. The basic idea is to sample the reference clock with multiple phases. If the DLL locks properly, assuming a duty-cycle of 0.5 for the reference clock, the first half of the phases

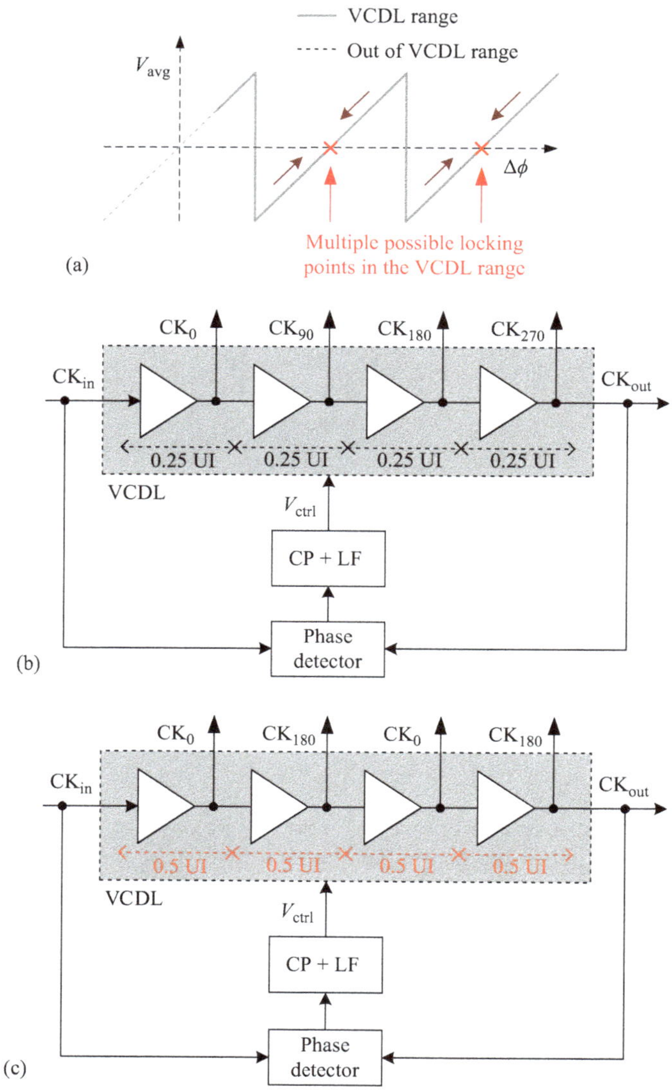

Figure 5.30 Harmonic locking of DLL

should sample "1" but the last half samples are "0". On the other hand, under the harmonic locking, the sampled pattern becomes different, so we are able to detect if the DLL falls into the harmonic locking or not.

Because a PLL corrects phase through adjusting frequency, the adjustable phase range is unlimited. On the other hand, a DLL has a finite range because it relies on the delay which can never be infinite or negative. That is, there is always a lower bound and an upper bound of the delay range. At start-up, if the initial delay starts from nearby those bounds, the phase detector drives the DLL to the locking point which is not in the available delay range, as shown in Figure 5.31. In this case, the phase detector tries to increase (or decrease) the delay while the delay is already set to the maximum (minimum), so the DLL is stuck at the bound. Because the stuck locking is a matter of the initial delay, a designer should carefully set the initial delay. In addition, a temperature drift or a voltage step may result in loss of lock and also affects the initial delay. As a result, continuous checking of the stuck state is also required. Generally, the stuck state is detected by detecting a condition where the VCDL control voltage (or code) is set to the maximum [5,29]. More strictly, the stuck state is where the phase detector tries to increase (decrease) the control voltage when it is already at the maximum [5], which can be detected by the stuck detector as shown in Figure 5.32.

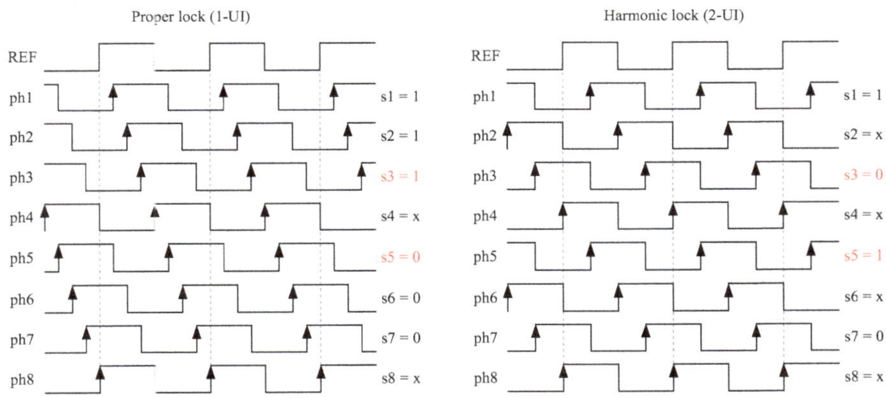

Figure 5.31 Harmonic locking detection technique based on multiphase sampling

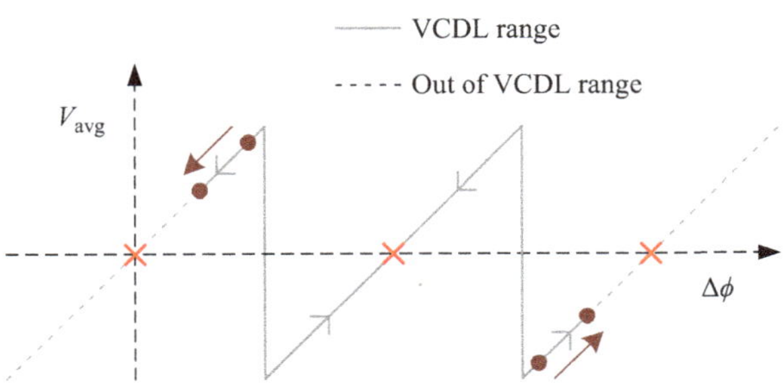

Figure 5.32 Illustration of stuck locking of DLL

Considering that the stuck locking is caused by the finite dynamic range of VCDL, using a phase interpolator as an infinite range VCDL can be a good approach depending on applications [30]. Taking multiphase clock from another PLL/DLL, a phase interpolator produces an intermediate phase by interpolating two of the phases (usually adjacent). Because of the rotating nature of phase, a phase interpolator can produce an infinite delay range once an adequate number of phases (typically no less than four) is provided, as shown in Figure 5.33(a). A circuit diagram of a digital phase interpolator, an example of phase interpolator implementations, is shown in Figure 5.33(b). It consists of two stages. The first stage is a MUX which selects two adjacent phases (P<*k*> and P<*k* + 1>) according to the coarse phase code. In the second stage, each path has a variable strength buffer whose weight (W_1 and W_2) is controlled by the fine phase code. The find code adjusts the ratio of W_1 and W_2 while the total weight is fixed. As a result, the output will be a blend of the P<*k*> weighted with W_1 and P<*k* + 1> weighted with W_2. Here, however, note that a linear combination of two clocks does not mean a linear combination of the two phases. The linearity of edge interpolation is a function of the slew rate of the interpolated signals. For example, if the slew rate is extremely fast, the edge cannot be interpolated as shown in the left diagram of Figure 5.33(c). With a lower slew rate, we can obtain a better linearity; however, the gradual clock edge is susceptible to the noise and leads to a high-power consumption due to the short-circuit current. The slew-rate susceptibility makes the phase interpolator design difficult due to the PVT variations. It becomes much worse if a wide frequency range must be supported. Note that because a lower slew rate leads to a better linearity, the phase interpolator typically shows a better performance at the slow PVT corner and higher input frequency. A slew-rate control can be used to retain the linearity across the PVT corners and the frequency range (Figure 5.34).

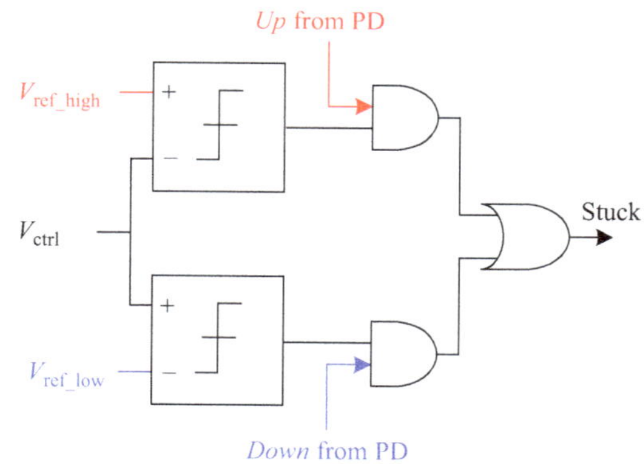

Figure 5.33 Stuck detector presented in [5]

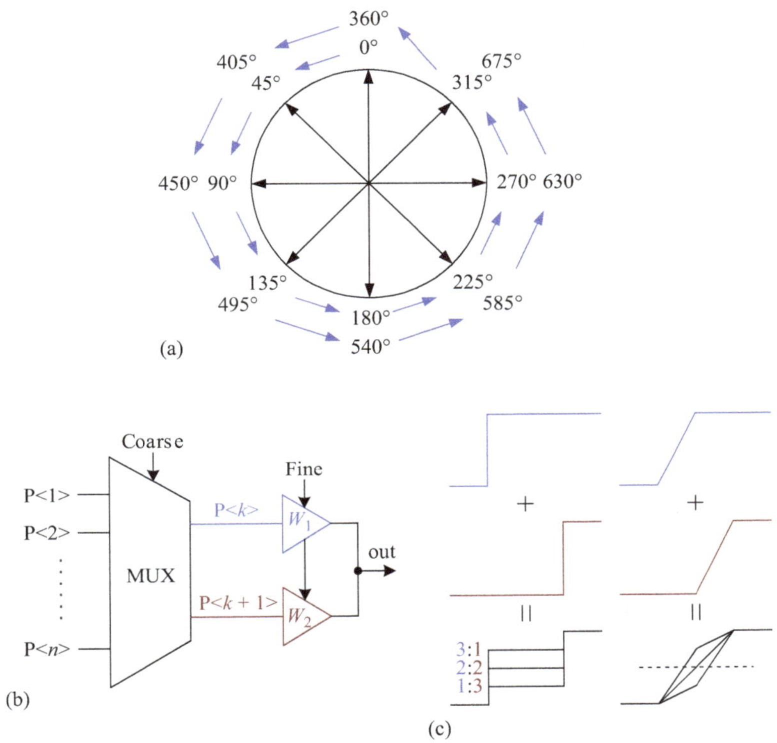

Figure 5.34 *(a) Infinite range of phase interpolator, (b) an example of phase interpolator implementation (digital phase interpolator), and (c) phase interpolator dependency on the slew rate*

References

[1] Frerking M. *Crystal Oscillator Design and Temperature Compensation*. New York: Springer Science & Business Media; 2012.

[2] Casper B, O'Mahony F. Clocking analysis, implementation and measurement techniques for high-speed data links: A tutorial. *IEEE Transactions on Circuits and Systems I: Regular Papers*. 2009;56(1):17–39.

[3] Jeong DK, Borriello G, Hodges DA, Katz RH. Design of PLL-based clock generation circuits. *IEEE Journal of Solid-State Circuits*. 1987;22(2): 255–261.

[4] Stojanovic V, Horowitz M. Modeling and analysis of high-speed links. In *Proceedings of the IEEE 2003 Custom Integrated Circuits Conference*. San Jose, CA: IEEE; 2003 (pp. 589–594).

[5] Bae W, Jeong GS, Park K, Cho SY, Kim Y, Jeong DK. A 0.36 pJ/bit, 0.025 mm^2, 12.5 Gb/s forwarded-clock receiver with a stuck-free delay-locked

loop and a half-bit delay line in 65-nm CMOS technology. *IEEE Transactions on Circuits and Systems I: Regular Papers*. 2016;63(9):1393–1403.

[6] Wu L, Black WC. A low-jitter skew-calibrated multi-phase clock generator for time-interleaved applications. In *2001 IEEE International Solid-State Circuits Conference, 2001. Digest of Technical Papers (ISSCC)*. San Francisco, CA: IEEE; 2001 (pp. 396–397).

[7] Lee K, Park J, Lee JW, *et al.* A single-chip 2.4-GHz direct-conversion CMOS receiver for wireless local loop using multiphase reduced frequency conversion technique. *IEEE Journal of Solid-State Circuits*. 2001;36(5):800–809.

[8] Lee MJ, Dally WJ, Chiang P. Low-power area-efficient high-speed I/O circuit techniques. *IEEE Journal of Solid-State Circuits*. 2000;35(11):1591–1599.

[9] Tang Y, Hu J, Park J, *et al.* A CMOS highly linear hybrid current/voltage controlled oscillator for wideband polar modulation. *IEEE Transactions on Circuits and Systems I: Regular Papers*. 2013;60(8):1991–2000.

[10] Andreani P, Mattisson S. On the use of MOS varactors in RF VCOs. *IEEE Journal of Solid-State Circuits*. 2000;35(6):905–910.

[11] Lo SH, Buchanan DA, Taur Y, Wang W. Quantum-mechanical modeling of electron tunneling current from the inversion layer of ultra-thin-oxide nMOSFET's. *IEEE Electron Device Letters*. 1997;18(5):209–211.

[12] Schmitz J, Cubaynes FN, Havens RJ, De Kort R, Scholten AJ, Tiemeijer LF. RF capacitance-voltage characterization of MOSFETs with high leakage dielectrics. *IEEE Electron Device Letters*. 2003;24(1):37–39.

[13] Moon Y, Choi J, Lee K, Jeong DK, Kim MK. An all-analog multiphase delay-locked loop using a replica delay line for wide-range operation and low-jitter performance. *IEEE Journal of Solid-State Circuits*. 2000;35(3):377–384.

[14] Walker RC. Designing bang-bang PLLs for clock and data recovery in serial data transmission systems. In *Phase-Locking in High-Performance Systems: From Devices to Architectures*. New York: Wiley-IEEE Press; 2003 (pp. 34–45).

[15] Hogge C. A self correcting clock recovery circuit. *Journal of Lightwave Technology*. 1985;3(6):1312–1314.

[16] Alexander JD. Clock recovery from random binary signals. *Electronics Letters*. 1975;11(22):541–542.

[17] Razavi B. Challenges in the design high-speed clock and data recovery circuits. *IEEE Communications Magazine*. 2002;40(8):94–101.

[18] Mansuri M, Liu D, Yang CK. Fast frequency acquisition phase-frequency detectors for Gsamples/s phase-locked loops. *IEEE Journal of Solid-State Circuits*. 2002;37(10):1331–1334.

[19] Perrott M. *High Speed Communication Circuits and Systems*. 2004. Available from: https://cppsim.com/CommCircuitLectures/lec14.pdf.

[20] Bae W, Ju H, Park K, Cho SY, Jeong DK. A 7.6 mW, 414 fs RMS-jitter 10 GHz phase-locked loop for a 40 Gb/s serial link transmitter based on a

two-stage ring oscillator in 65 nm CMOS. *IEEE Journal of Solid-State Circuits*. 2016;51(10):2357–2367.

[21] Razavi B. TSPC logic [a circuit for all seasons]. *IEEE Solid-State Circuits Magazine*. 2016;8(4):10–13.

[22] Riley TA, Copeland MA, Kwasniewski TA. Delta-sigma modulation in fractional-N frequency synthesis. *IEEE Journal of Solid-State Circuits*. 1993;28(5):553–559.

[23] Hsu CM, Straayer MZ, Perrott MH. A low-noise wide-BW 3.6-GHz digital $\Delta\Sigma$ fractional-n frequency synthesizer with a noise-shaping time-to-digital converter and quantization noise cancellation. *IEEE Journal of Solid-State Circuits*. 2008;43(12):2776–2786.

[24] Jones KD, Kim J, Konrad V. Some "real world" problems in the analog and mixed signal domains. *Proceedings of Designing Correct Circuits*. 2008 (pp. 51–68). Available from: http://www.cs.um.edu.mt/gordon.pace/Workshops/DCC2008/Presentations/10.pdf.

[25] Greenstreet MR, Yang S. Verifying start-up conditions for a ring oscillator. In *Proceedings of the 18th ACM Great Lakes Symposium on VLSI*. Orlando, FL: ACM; 2008 (pp. 201–206).

[26] Kim T, Song DG, Youn S, Park J, Park H, Kim J. Verifying start-up failures in coupled ring oscillators in presence of variability using predictive global optimization. In *Proceedings of the International Conference on Computer-Aided Design*. San Jose, CA: IEEE Press; 2013 (pp. 486–493).

[27] Jung YJ, Lee SW, Shim D, Kim W, Kim C, Cho SI. A dual-loop delay-locked loop using multiple voltage-controlled delay lines. *IEEE Journal of Solid-State Circuits*. 2001;36(5):784–791.

[28] Byun S, Park CH, Song Y, Wang S, Conroy CS, Kim B. A low-power CMOS Bluetooth RF transceiver with a digital offset canceling DLL-based GFSK demodulator. *IEEE Journal of Solid-State Circuits*. 2003;38(10): 1609–1618.

[29] Yang RJ, Liu SI. A 40–550 MHz harmonic-free all-digital delay-locked loop using a variable SAR algorithm. *IEEE Journal of Solid-State Circuits*. 2007;42(2):361–373.

[30] Sidiropoulos S, Horowitz MA. A semidigital dual delay-locked loop. *IEEE Journal of Solid-State Circuits*. 1997;32(11):1683–1692.

Chapter 6
PLL loop dynamics and jitter

6.1 Transfer function of PLL building blocks

In this chapter, the phase domain transfer function of each building block of the PLL is described. Because the intent of the PLL is "phase lock," the analysis should be done in the phase domain, so it is assumed that a phase error (ϕ_{err}) is applied to the input of the PLL. For the derivation of the loop dynamics, ϕ_{err} is assumed to be small enough and to be introduced after the PLL achieves the phase lock. Recalling Chapter 5, the PFD converts the ϕ_{err} to the voltage averaged over the reference period (V_{avg}), as shown in Figure 6.1. Note that what the PFD actually produces is UP and DN pulses whose pulse-width difference is proportional to the ϕ_{err}. The average voltage is obtained by assigning the weight of $+V_{max}$ and $-V_{max}$ to the UP and DN, respectively, where V_{max} corresponds to the voltage of logical "high," which typically equals V_{DD}. Therefore, the V_{avg} can be obtained from ϕ_{err} as

$$V_{avg} = \frac{V_{max}}{2\pi}\phi_{err} \tag{6.1}$$

The charge pump flows the pull-up (or pull-down) current I_{CP} only when the UP (or DN) pulse is high, the average current can be written as

$$I_{avg} = \frac{V_{avg}}{V_{max}}I_{CP} = \frac{I_{CP}}{2\pi}\phi_{err} \tag{6.2}$$

which leads that the combined transfer function of the PFD and charge pump is $I_{CP}/2\pi$ (A/rad). The current is transformed to the control voltage (V_{ctrl}) for the VCO by the loop filter. The current flows to the ground through the loop filter impedance, so the V_{ctrl} is a simple multiple of the loop filter impedance and the average current. As a result, the deviation of V_{ctrl} over the equilibrium voltage introduced by the phase error is written as

$$\Delta V_{ctrl} = \frac{I_{CP}Z_{LF}(s)}{2\pi}\phi_{err} \tag{6.3}$$

We need to be careful when we derive the transfer function of the VCO and the frequency divider because they are less straightforward compared to those of the PFD, charge pump, and loop filter. As defined in Chapter 5, the VCO output frequency is a function of the V_{ctrl} and the slope of the frequency-voltage curve is

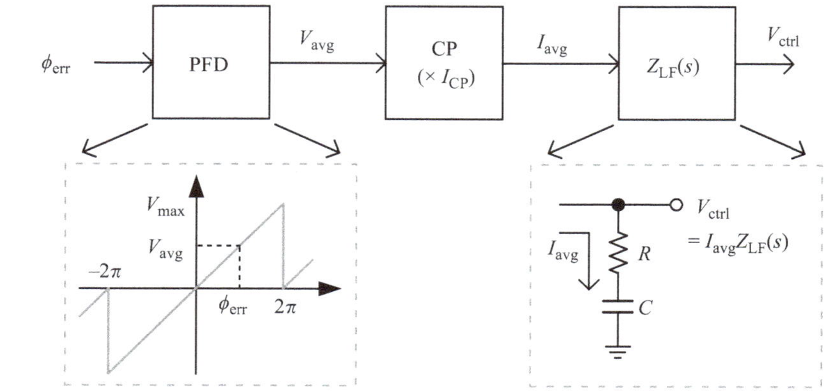

Figure 6.1 Transfer function of PFD, charge pump, and loop filter

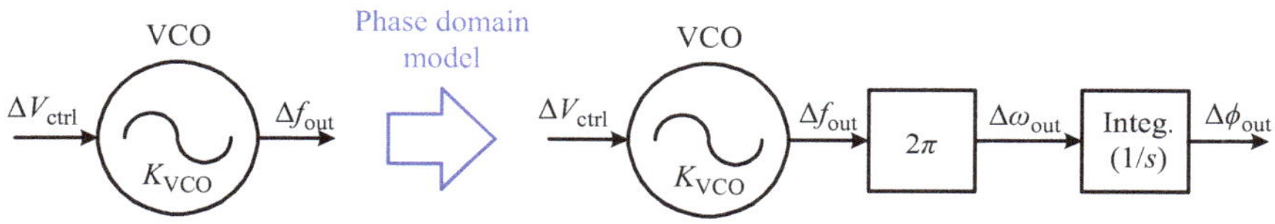

Figure 6.2 Transfer function of VCO

defined as K_{VCO}. Either of rad/s/V (for angular frequency) and Hz/V (for frequency) can be used for the unit of K_{VCO}; however, they cause a little difference to the transfer function. For clarity, in the following derivations, we will use Hz/V because the frequency is the preferred unit for the CMOS clocking circuit. Therefore, the frequency shift by the ΔV_{ctrl} is simply written as

$$\Delta f_{out} = K_{VCO}\Delta V_{ctrl} \tag{6.4}$$

However, as mentioned above the derivation should be obtained in the phase domain, the frequency shift should be converted to the phase shift. Therefore, the K_{VCO} should be followed by an integrator to obtain the phase shift so that the transfer function of the VCO becomes:

$$\frac{\Delta\phi_{out}}{\Delta V_{ctrl}} = \frac{2\pi K_{VCO}}{s} \tag{6.5}$$

Note that 2π is introduced to convert the frequency to the angular frequency since we used Hz/V for the unit of the K_{VCO} (Figure 6.2).

When we recall the operation of the frequency divider, the division ratio is determined by how many edges of the input clock are required to trigger the output, from which we can find that the output edge is synchronized to the input edge if we neglect the additive noise. That is, a time error (ΔT_{in}) of the input is preserved at the

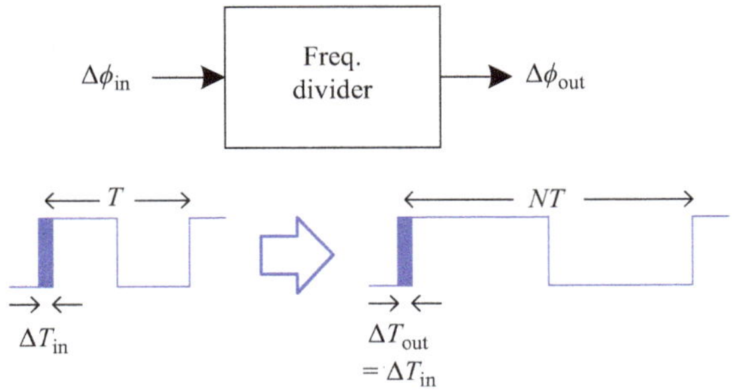

Figure 6.3 Transfer function of frequency divider

output while the period becomes N times longer. Therefore, the output phase is obtained as

$$\Delta\phi_{out} = \frac{2\pi\Delta T_{out}}{NT} = \frac{1}{N}\cdot\frac{2\pi\Delta T_{in}}{T} = \frac{1}{N}\cdot\Delta\phi_{in} \tag{6.6}$$

which leads the transfer function of the frequency divider to $\frac{1}{N}$ (Figure 6.3).

6.2 PLL loop dynamics

6.2.1 Second-order PLL

For a negative feedback system, the closed-loop transfer function is given as

$$H(s) = \frac{T(s)}{1 + kT(s)} \tag{6.7}$$

where $T(s)$ is the open-loop transfer function and k is the feedback factor, which is $1/N$ in the PLL. Because the open-loop gain at low frequency is infinite due to the integration, the DC gain of the closed-loop equals N. Therefore, the cut-off bandwidth (ω_c, or 3-dB bandwidth) of the closed-loop transfer function can be obtained by equating:

$$|H(\omega_c)|^2 = \frac{|T(\omega_c)|^2}{\left(1 + \left|\frac{T(\omega_c)}{N}\right|\cos(\angle T(\omega_c))\right)^2 + \left(\left|\frac{T(\omega_c)}{N}\right|\sin(\angle T(\omega_c))\right)^2} = \frac{N^2}{2} \tag{6.8}$$

which leads to

$$|T(\omega_c)| = N\left(\cos(\angle T(\omega_c)) + \sqrt{\cos^2(\angle T(\omega_c)) + 1}\right) \tag{6.9}$$

Assuming a large $\angle T(\omega_c)$ (~90°), (6.9) can be simplified to

$$|T(\omega_c)| \cong N \tag{6.10a}$$

but with a moderate $\angle T(\omega_c)$ (~45°)

$$|T(\omega_c)| \cong 2N \tag{6.10b}$$

Combining the transfer functions of the building blocks, the open-loop transfer function of the second-order PLL is written as

$$T(s) = \frac{I_{\mathrm{CP}}}{2\pi}\left(R + \frac{1}{sC}\right)\frac{2\pi K_{\mathrm{VCO}}}{s} = \frac{(1 + sRC)I_{\mathrm{CP}}K_{\mathrm{VCO}}}{s^2 C} \tag{6.11}$$

Figure 6.4 illustrates the Bode plot of (6.11). Because the $T(s)$ has two poles at DC, the phase response starts from $-180°$, and the slope of the magnitude response is -40 dB/dec at a low frequency. The phase response gradually increases from $-180°$ and becomes $-135°$ at the ω_z. Notice that here we assume that ω_c is higher than ω_z. Recalling from (3.9), a negative feedback system becomes an oscillator when the magnitude of the open-loop gain is unity while the phase response equals $-180°$ (or 180°). From this observation, we can define a metric "phase margin" as how far the phase is apart from $-180°$ at ω_c, which shows how stable the feedback system is

$$\mathrm{PM} = \angle T(\omega_c) + 180° \tag{6.12}$$

The phase margin is directly coupled to the damping factor (ζ) of the step response of the system, which is given as

$$\mathrm{PM} = \arctan\left(\frac{2\zeta}{\sqrt{-2\zeta^2 + \sqrt{1 + 4\zeta^4}}}\right) \tag{6.13}$$

In general, a reasonable range of the damping factor is $0.5 < \zeta < 2.0$, which approximately corresponds to $45° < \mathrm{PM} < 85°$, because low damping factor leads to an unstable ringing whereas high damping ratio makes the step response be slow. Moreover, a jitter peaking becomes significant at low and high damping factors [1]. In a practical implementation, the higher end of the PM range is reduced to around 70°, because a larger loop filter capacitor is required for achieving a higher PM.

From (6.10) and (6.11), the cut-off frequency is calculated by equating:

$$|T(\omega_c)| = \frac{(1 + \omega_c RC)I_{\mathrm{CP}}K_{\mathrm{VCO}}}{\omega_c^2 C} \cong 2N \tag{6.14}$$

Assuming $\omega_c \gg \frac{1}{RC}$, we can get:

$$\omega_c \cong \frac{I_{\mathrm{CP}}RK_{\mathrm{VCO}}}{2N} \tag{6.15}$$

$$|AH(j\omega_0)| = 1,\ \angle AH(j\omega_0) = 2 \tag{3.9}$$

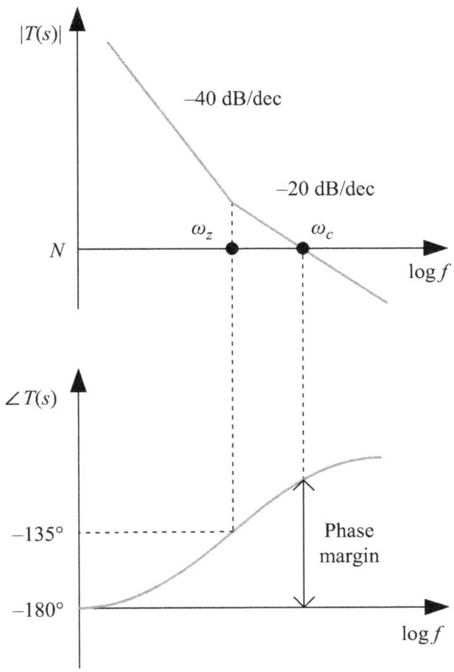

Figure 6.4 Bode plot of the second-order PLL open-loop transfer function

On the other hand, the phase response of the PLL can be written as

$$\angle T(\omega) = -180^\circ + \arctan(RC\omega) \tag{6.16}$$

The phase margin of the loop is obtained from (6.12) as

$$\mathrm{PM} = \arctan(RC\omega_c) = \arctan\left(\frac{\omega_c}{\omega_z}\right) \tag{6.17}$$

which implies that the assumption of $\omega_c \gg \frac{1}{RC}$ is valid when the loop has an enough phase margin.

From another point of view, we can also calculate how much a PLL responses to an input phase error in the time domain, which can be more straightforward than the frequency domain. Note that the phase error is simply converted to the time error as

$$T_{\text{err}} = \frac{\phi_{\text{err}}}{2\pi} \cdot \frac{1}{f_{\text{ref}}} = \frac{\phi_{\text{err}}}{\omega_{\text{ref}}} \tag{6.18}$$

Let us assume that the frequency is locked and neglects the frequency tracking by the loop filter capacitor for simplicity. Then a voltage pulse by the IR drop across the resistor raises on the top of the PLL's steady-state control voltage. The amount of updated phase at the output of the VCO due to the control voltage pulse becomes:

$$\Delta\phi_{\mathrm{out}} = \frac{T_{\mathrm{err}}}{T_{\mathrm{ref}}} \cdot I_{\mathrm{CP}} \cdot R \cdot \frac{2\pi K_{\mathrm{VCO}}}{\omega_{\mathrm{VCO}}} = \frac{I_{\mathrm{CP}} R K_{\mathrm{VCO}}}{N} \cdot T_{\mathrm{err}} \tag{6.19}$$

Substituting (6.15) and (6.18) into (6.19), the PLL response to the phase error becomes:

$$\frac{\Delta\phi_{\mathrm{out}}}{\phi_{\mathrm{err}}} \cong \frac{2\omega_c}{\omega_{\mathrm{ref}}} = \frac{2f_c}{f_{\mathrm{ref}}} \tag{6.20}$$

Equation (6.20) gives an interesting insight of the PLL; the 3-dB bandwidth over the reference frequency is linearly proportional to the ratio of $\Delta\phi_{\mathrm{out}}$ over ϕ_{err}, that is, how much the output phase responds to the input phase difference.

From above, it is assumed that the PLL does not have a loop latency. However, the PLL is actually a discrete-time system so there must be a loop latency. Including the latency, the open-loop transfer function becomes:

$$T(s) = \frac{(1 + sRC)I_{\mathrm{CP}}K_{\mathrm{VCO}}}{s^2 C} \mathrm{e}^{-s\tau} \tag{6.21}$$

where τ is the loop latency. Considering the PLL as a discrete-time system, the sampling rate is the reference frequency. As a result, if a phase error is detected at a certain cycle, the response to the error will appear at the next cycle at the earliest. From that, we can find that the latency can be replaced by $\frac{1}{\omega_{\mathrm{ref}}}$. With the inclusion of the latency, the phase response and the phase margin of the PLL are rewritten as

$$\begin{aligned} \angle T(\omega_c) &= -180^\circ + \arctan\left(\frac{\omega_c}{\omega_z}\right) - \omega_c \tau \cdot \left(\frac{180^\circ}{\pi}\right) \\ &= -180^\circ + \arctan\left(\frac{\omega_c}{\omega_z}\right) - \frac{\omega_c}{\omega_{\mathrm{ref}}} \cdot \left(\frac{180^\circ}{\pi}\right) \end{aligned} \tag{6.22}$$

$$\mathrm{PM} = \arctan\left(\frac{\omega_c}{\omega_z}\right) - \frac{\omega_c}{\omega_{\mathrm{ref}}} \cdot \left(\frac{180^\circ}{\pi}\right) \tag{6.23}$$

It is found that there is a phase-margin-degrading factor caused from the latency. Without considering the discrete-time effect, the phase margin increases as the cut-off frequency increases. However, the latency introduces the degrading factor which is linearly proportional to the cut-off frequency. Note that the slope of arctan(x) is steep for a small x but becomes gradual for a higher x. As a result, for a small cut-off frequency the arctangent term dominates so that the phase margin increases as the cut-off frequency increases; however, the degrading term starts dominating the slope so that the phase margin decreases as the cut-off frequency increases. This trend can be found in Figure 6.5, where ω_z is 1/50 of ω_{ref}. As a rule of thumb, it is recommended to have the cut-off frequency lower than 1/10 (sometimes 1/15) of the reference

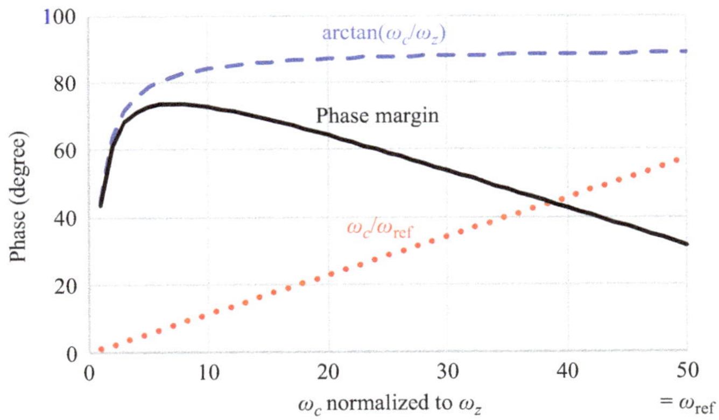

Figure 6.5 PLL phase margin considering the loop latency

frequency [2]. Note that the latency becomes even longer than $\frac{1}{\omega_{\text{ref}}}$ if the loop component (i.e., loop filter and divider) introduces latency which is longer than 1 UI.

6.2.2 Tuning design parameters

From Section 6.2.1, we derived the PLL loop dynamics. From now on, the impact of tuning each design parameter on the PLL dynamics will be studied. PLL designers must understand which design parameters they should deal with, and how they should be adjusted. From (6.11), the open-loop transfer function can be approximated as

$$T(s) = \frac{I_{\text{CP}} K_{\text{VCO}}}{s^2 C} \quad \left(\omega < \frac{1}{RC}\right) \tag{6.24}$$

$$T(s) = \frac{R I_{\text{CP}} K_{\text{VCO}}}{s} \quad \left(\omega > \frac{1}{RC}\right) \tag{6.25}$$

which is illustrated in Figure 6.6(a). Note that (6.24) and (6.25) correspond to −40 dB/dec region and −20 dB/dec region of Figure 6.4, respectively. Equations (6.24) and (6.25) imply that tuning the loop filter capacitance affects the −40 dB/dec line whereas the −20 dB/dec line keeps the same regardless of the capacitance. Using smaller capacitor shift the line to the positive direction, which means ω_z becomes closer to ω_c so that the phase margin is degraded, according to (6.23). Considering that a smaller capacitance results in a bigger frequency shift from the same input stimulus, the physical implication of this observation is that a relatively higher gain at the frequency control (integral control) degrades the loop stability. Intuitively, a larger phase control gain is preferable for a better phase-locking, but a larger frequency control gain is beneficial for a frequency-locking purpose. As a result, larger capacitance provides a better stability during the normal operation but also results in a long frequency acquisition time from start-up. Back to Figure 6.6, note that ω_c is not affected by the capacitance as long as ω_z is lower

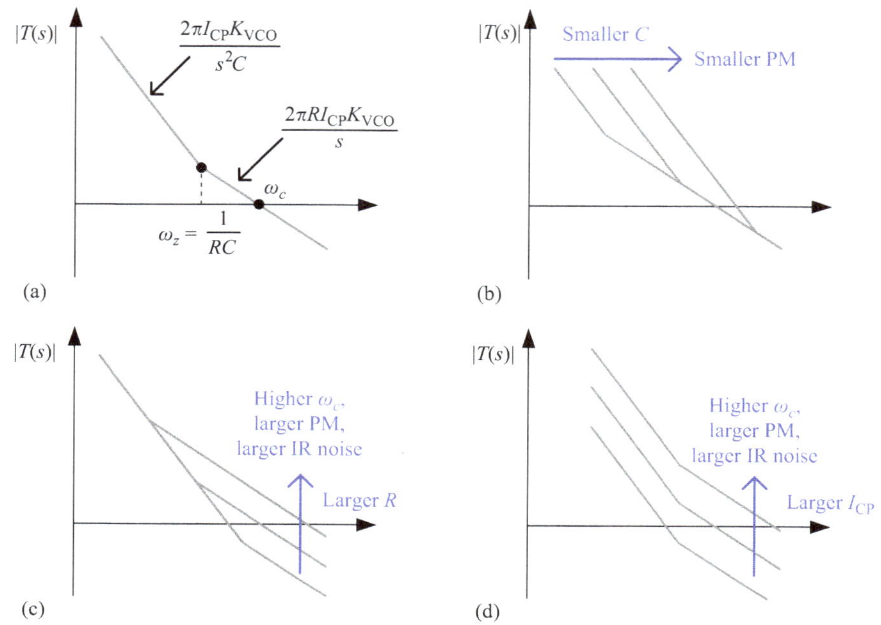

Figure 6.6 (a) Approximation of the open-loop gain, (b) PLL dynamics variation by capacitor tuning, (c) PLL dynamics variation by resistor tuning, (d) and PLL dynamics variation by charge pump current tuning

than ω_c. If ω_z is higher than ω_c, ω_c becomes higher as the capacitance decreases, but this case is generally out of interest because the loop is not stable. On the other hand, tuning the resistance shifts −20 dB/dec line since only (6.25) depends on the resistance. Therefore, increasing the resistance lowers ω_z but raises ω_c, thus the phase margin increases steeper than the capacitive tuning as far as ω_c is lower than $\frac{\omega_{ref}}{10}$. On the downside, the larger resistance also increases an instant IR drop due to the non-ideality of the PFD, loop filter, and charge-pump. It leads to a reference spur which refers to a spurious spectral component at the reference frequency. The reference spur will be studied in Section 6.2.4. On the other hand, tuning the charge-pump current affects both the −40 dB/dec region and the −20 dB/dec region. Therefore, increasing the charge-pump current shifts the entire curve to the positive direction. The ω_c increases as the current increases hence the phase margin also increases; however, the ω_z stays constant in contrast to the resistor tuning case. That means the amount of phase margin enhancement is lower than that with the resistor tuning. For instance, doubling the pump current increases the phase margin to 49° from 30°, while doubling the resistance increases the phase margin to 66°.

6.2.3 PLL jitter

Every loop component, we went through in Section 5.2, is a source of jitter. Figure 6.7 illustrates the jitter sources in a PLL: jitter of the reference clock ($\Delta\phi_{ref}$),

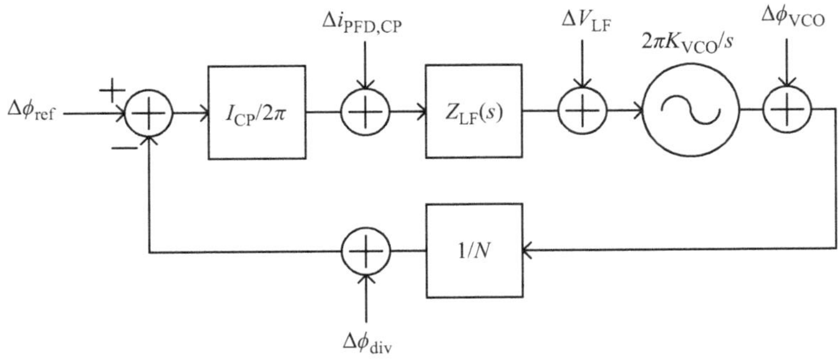

Figure 6.7 Jitter sources in PLL

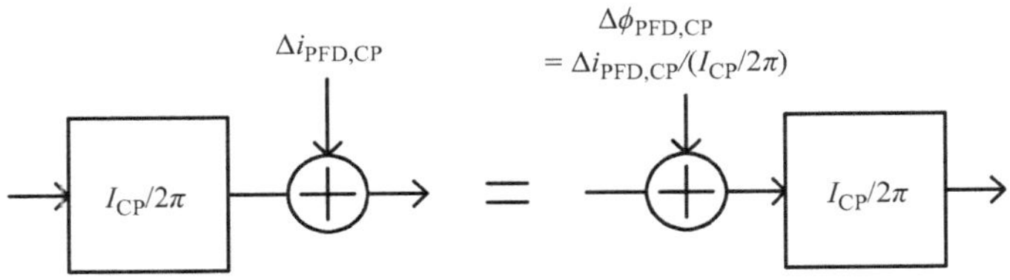

Figure 6.8 PFD-CP noise conversion to input-referred jitter

the PFD and charge-pump noise ($\Delta I_{PFD,CP}$), jitter added by the divider ($\Delta\phi_{div}$), the loop filter noise (ΔV_{LF}), and the VCO phase jitter ($\Delta\phi_{VCO}$). For the given noise contribution from the noise sources, PLL output jitter can be suppressed through a proper selection of loop dynamics, that is, bandwidth. Therefore, it is important to understand the jitter transfer function of each sources so that we can design an optimal loop that suppresses the overall jitter to be the minimum. At first, we already derived the transfer function from the input reference clock to the PLL output clock in (6.7) and (6.11), which implies that the input jitter is low-pass-filtered with the cut-off frequency of ω_c (6.15). It is also easy to find that $\Delta\phi_{div}$ is injected to the loop at the same point as $\Delta\phi_{ref}$, thus the jitter transfer function of $\Delta\phi_{div}$ equals to that of $\Delta\phi_{ref}$.

The PFD and charge-pump noise, which is represented to an additive current noise at the output of the charge-pump, also exhibit a similar behavior as $\Delta\phi_{ref}$. Because the transfer function of the PFD and charge-pump is independent to the frequency, the current noise can be transformed to the input-referred jitter simply by being divided by $I_{CP}/2\pi$, as shown in Figure 6.8. After transforming, the input referred jitter occurs at the same point that the $\Delta\phi_{ref}$ is injected to the loop so that it has the same low-pass filter transfer function.

On the other hand, the transfer function becomes different for the ΔV_{LF} and the $\Delta\phi_{VCO}$, because they are injected at a different point of the loop. Figure 6.9 shows the derivation process of the ΔV_{LF} transfer function to the output jitter. The output phase deviation ($\Delta\phi_{out}$) turns around through the feedback path and is added to the

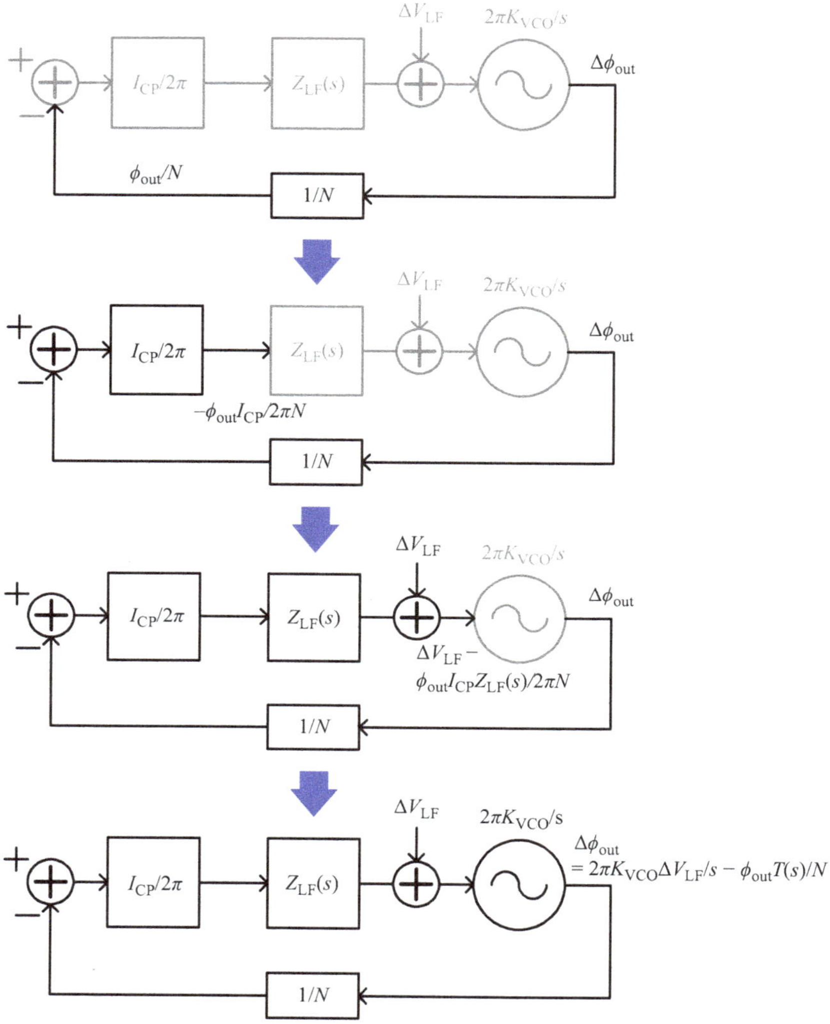

Figure 6.9 Derivation of jitter transfer function of loop-filter noise

ΔV_{LF} after passing the divider, PFD, and charge-pump. The added voltage is fed to the VCO, and therefore the $\Delta\phi_{out}$ can be written as

$$\Delta\phi_{out} = \frac{2\pi K_{VCO}\Delta V_{LF}}{s} - \frac{T(s)}{N}\Delta\phi_{out} \tag{6.26}$$

By equating (6.26), we can obtain the transfer function of the ΔV_{LF} as

$$\frac{\Delta\phi_{out}}{\Delta V_{LF}} = \frac{2\pi K_{VCO}}{s\left(1+\frac{T(s)}{N}\right)} = \frac{2\pi K_{VCO}s}{s^2 + \frac{(sRC+1)I_{CP}K_{VCO}}{NC}} = \frac{2\pi K_{VCO}s}{s^2 + \frac{\omega_c}{\sqrt{2}N}s + \frac{\omega_c\omega_z}{\sqrt{2}N}} \tag{6.27}$$

which is a band-pass filter whose peak frequency is generally below the ω_z. Since the thermal noise of the resistor is the dominant source of the loop filter noise, the jitter added by the loop filter can be written as

$$\Delta\phi^2_{\text{out,LF}} = 4kTR \cdot \left(\frac{2\pi K_{\text{VCO}} s}{s^2 + \frac{\omega_c}{\sqrt{2N}} s + \frac{\omega_c \omega_z}{\sqrt{2N}}} \right)^2 \tag{6.28}$$

Note that the shape of (6.27) is a complex function of ω_c, ω_z, and N. Figure 6.10 visualizes the magnitude response of (6.27) while tuning N and ω_c/ω_z. The magnitude and the frequency are normalized to $2\pi K_{\text{VCO}}$ and ω_z, respectively. We can find that the more loop filter noise is filtered for a smaller N if the same resistance is assumed. The filtering capability also increases as ω_c/ω_z increases. Recall that higher ω_c/ω_z corresponds to a higher phase margin.

Since correcting the frequency instability of an oscillator is one of the primary motives of the PLL, it is very important for a PLL designer to understand how $\Delta\phi_{\text{VCO}}$ is filtered out in the PLL. Figure 6.11 explains the derivation of the jitter transfer of the $\Delta\phi_{\text{VCO}}$. Similar to the derivation of the ΔV_{LF} transfer, the $\Delta\phi_{\text{out}}$ travels through the loop but the $\Delta\phi_{\text{VCO}}$ is added at the output of the VCO. Therefore, the $\Delta\phi_{\text{VCO}}$ jitter transfer is obtained by equating:

$$\Delta\phi_{\text{out}} = \Delta\phi_{\text{VCO}} - \frac{T(s)}{N} \Delta\phi_{\text{out}} \tag{6.29}$$

Equation (6.29) leads to

$$\frac{\Delta\phi_{\text{out}}}{\Delta\phi_{\text{VCO}}} = \frac{1}{1 + \frac{T(s)}{N}} \tag{6.30}$$

which is a high-pass filter whose cut-off frequency is ω_c.

Equation (6.30) implies an important criterion of PLL design. The VCO phase noise is high-pass-filtered whereas the other noises (except for the loop filter noise) is low-pass filtered, and the cut-off frequency equals to ω_c for both cases. That is,

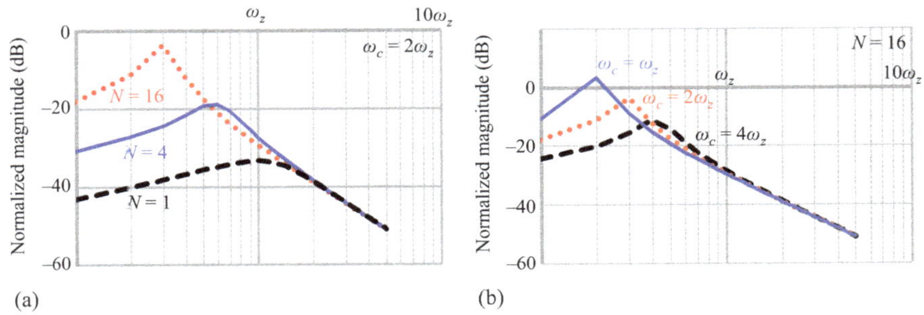

Figure 6.10 Trend of jitter transfer function of loop-filter noise with (a) tuning N and (b) tuning ω_c/ω_z

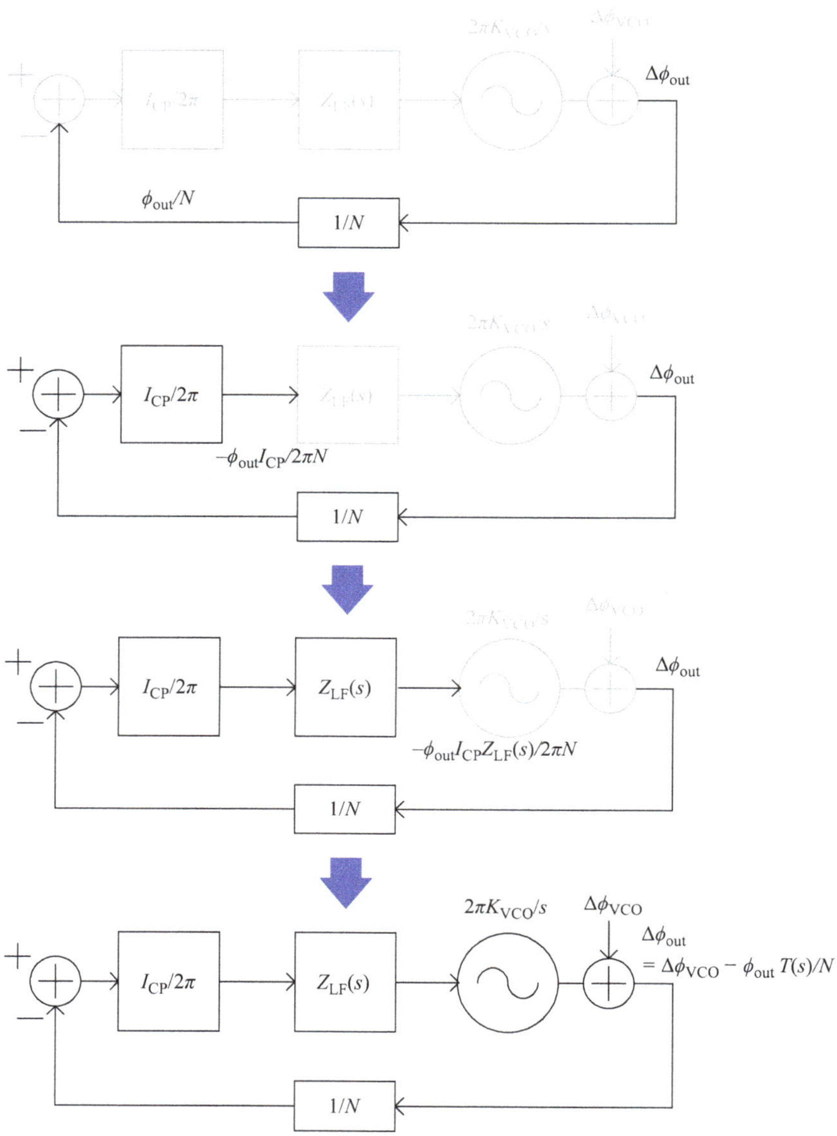

Figure 6.11 Derivation of jitter transfer function of VCO phase noise

the VCO phase noise is suppressed by a higher ω_c, whereas the input phase noise is less filtered out. What we can find from this trade-off is that there should be an optimum loop bandwidth that minimized the overall phase noise of a PLL. Figure 6.12 illustrates the trade-off. For simplicity, an ideal low-pass and high-pass filters with the cut-off frequency at the ω_c is assumed, and the loop filter noise

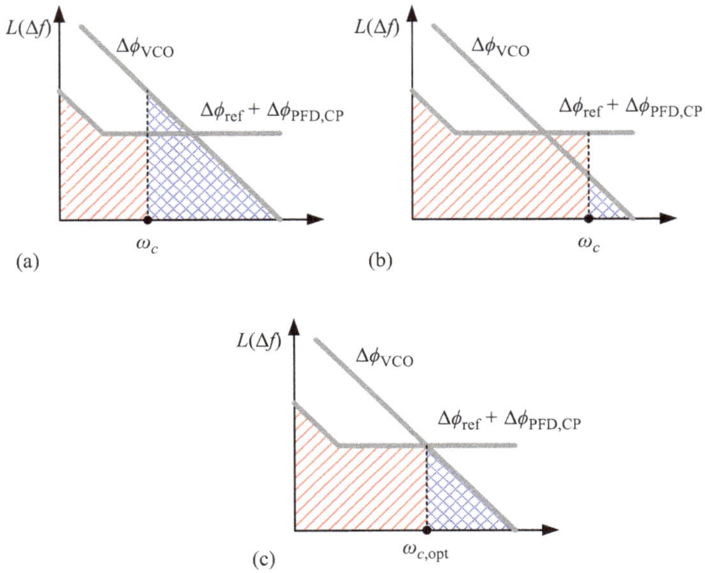

Figure 6.12 Relation between PLL overall phase noise and loop bandwidth

contribution is neglected. As observed in Figure 6.12(c), the optimum ω_c should be at the crossing point of the VCO phase noise and the input phase noise. If the ω_c is too low, the VCO phase noise is not sufficiently filtered out, so the VCO phase noise at the vicinity of the ω_c is excessively added to the optimum phase noise, which is illustrated in Figure 6.12(a). On the other hand, if the ω_c is too high then the excess phase noise is added from the phase noise from the input.

6.2.4 Reference spur and static phase error

The reference spur refers to the spurious tone observed in the PSD measurement of the PLL clock, at the frequency that is apart from the center frequency by the reference frequency (Figure 6.13). It is mainly due to the fact that the PLL operation is based on the periodic sampling by the PFD, whose sampling frequency equals the reference clock frequency. Many static nonidealities introduced from practical CMOS circuit designs would result in a fluctuation in the PLL control voltage, which is directly transformed to the fluctuation of the PLL clock frequency by K_{VCO} [3,4]. The fluctuation should be periodic at the rate of the reference frequency because of the sampling nature of the PLL. This section introduces major sources of the reference spur (leakage current of the loop filter capacitor, charge-pump mismatch between pull-up and pull-down currents, and PFD mismatch) and also describes the mechanisms how they cause the reference spur.

The first example of the non-ideality that causes the reference spur is the leakage current of the loop filter capacitance. MOS capacitor is frequently used to implement the PLL loop filter capacitor since it generally provides the highest

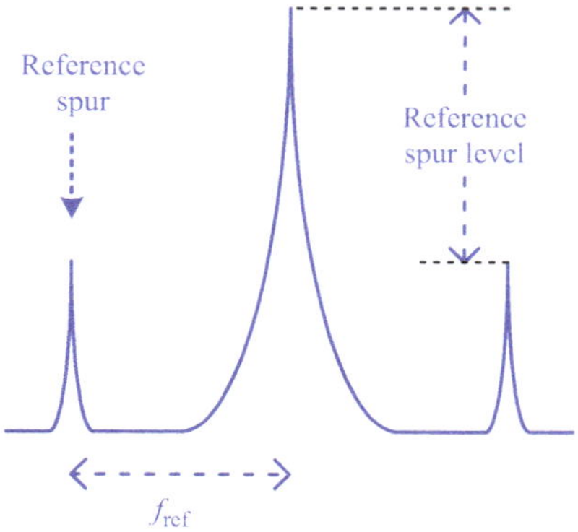

Figure 6.13 Reference spur illustrated in PSD domain

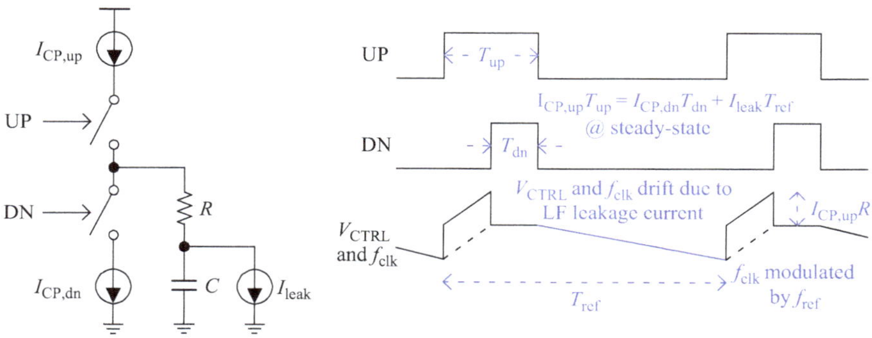

Figure 6.14 Reference spur due to loop filter leakage current

capacitance density in CMOS technology. However, in deep-submicron CMOS technology, a leakage current flows from the gate to the channel because a direct tunneling is occurred across the thinned gate oxide [5]. Figure 6.14 shows how the leakage current results in the reference spur. In a steady-state, the average frequency of the VCO clock should be the same as the reference frequency, which corresponds that the average of V_{ctrl} stays constant over the reference period as well. That means the net charge by the charging and discharging current should be zero. If the charge-pump UP/DN currents are matched, the UP and DN pulses have the same pulse-width. However, if a nonzero leakage current flows through the capacitor and discharges the V_{ctrl}, additional charges should be provided to nullify the charge loss by the leakage current. As a result, the UP pulse becomes wider than the DN pulse, which is enabled by introducing a static phase error of $(T_{up} - T_{dn})$ between the reference clock and the VCO feedback clock. Note that the UP and DN pulses are reset at the same time, if we neglect any mismatch in the PFD reset path, so the UP pulse is triggered earlier than the DN pulse. Let us look into the corresponding V_{ctrl} waveform. Once the rising edge of the reference clock is triggered, the UP pulse becomes high, and the V_{ctrl} raises. After the DN pulse is triggered to high, the V_{ctrl} stays because the UP and DN currents are canceled out. Once the PFD is reset, the leakage current starts discharging the capacitor until the next rising edge of the reference clock is triggered. We can find that the V_{ctrl} has a periodic waveform whose frequency is the same as the reference frequency. Assuming a linear transformation from V_{ctrl}, the VCO frequency is modulated at the reference frequency so that the PSD will show a spurious tone at the frequencies that are apart from the center frequency by the reference frequency.

The pull-up and pull-down current mismatch of the charge-pump is another major source of the reference spur. There are many nonidealities that cause the mismatch such as the finite output resistance of MOSFET device, transient discrepancy due to the inherent PMOS/NMOS mismatch, and device random mismatch. As shown in Figure 6.15, if the pump-current mismatch exists, the PLL tries to nullify the amount of charges injected to the loop filter by forcing the average PFD output to be non-zero, that is,

$$I_{CP,up}T_{up} = I_{CP,dn}T_{dn} \tag{6.31}$$

where $I_{CP,up}$, $I_{CP,dn}$, T_{up}, and T_{dn} denote the pull-up current, the pull-down current, the pulse-width of the UP signal, and the pulse-width of the DN signal, respectively. Again, note that the triggered UP and DN pulses are reset to zero at the same time if we neglect mismatch in the PFD circuit. That is if the $I_{CP,up}$ is smaller than $I_{CP,dn}$, the UP pulse is triggered earlier than the DN signal. Therefore, the control voltage raises when the up is high while the DN stays low. However, once the DN is triggered, the control voltage starts decreasing since $I_{CP,dn}$ is larger. The resulting waveform of the control voltage includes periodic ripples as shown in Figure 6.15, which can be observed as the reference spur in the PSD domain. Note that the pump current mismatch forces the static phase error of $T_{up} - T_{dn}$ as well.

The last example of the sources of the reference spur is the reset path mismatch in the PFD. Typically, the contribution from the PFD is less significant than that

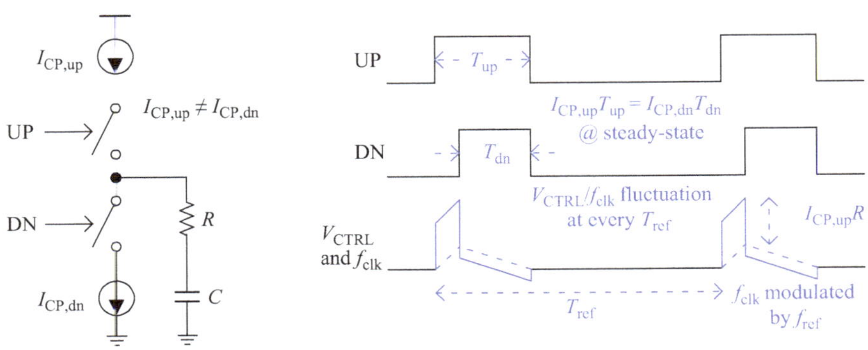

Figure 6.15 Reference spur due to charge-pump up/down current mismatch

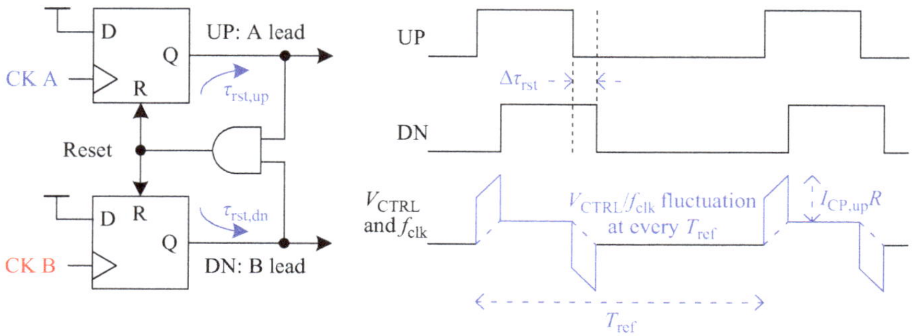

Figure 6.16 Reference spur due to PFD mismatch

from the loop filter and the charge-pump. Although both DFFs share the same AND gate, the reset path in each DFF would have a random mismatch. If the reset delays are different, the UP and DN pulses are shifted by the delay difference. The pulse shift introduces positive and negative pulses on the control voltage, whose amplitude is proportional to the amount of the phase shift, as shown in Figure 6.16. In this section, we looked over the reference spur qualitatively. A detailed derivation of analytic reference spur is given in Section 6.4.

6.2.5 Third-order PLL

A third-order PLL with an additional pole is widely used to suppress the reference spur. Using an additional parallel loop filter capacitor C_2 introduces an additional pole to the PLL transfer function as shown in Figure 6.17; therefore, the PLL becomes third-order. The impedance of the loop filter of the third-order PLL is calculated as

$$Z_{LF,3rd}(s) = \left(R + \frac{1}{sC_1}\right) \Bigg\| \frac{1}{sC_2} = \frac{RC_1 s + 1}{RC_1C_2s^2 + (C_1 + C_2)s}$$
$$= \frac{\frac{s}{\omega_z} + 1}{(C_1 + C_2)s} \cdot \frac{1}{\frac{s}{\omega_p} + 1} \tag{6.32}$$

where

$$\omega_z = \frac{1}{RC_1} \tag{6.33}$$

$$\omega_p = \frac{C_1 + C_2}{RC_1C_2} \tag{6.34}$$

To simplify (6.32), we can make an assumption. Adding a pole to the transfer function introduces an additional phase shift, which degrades the phase margin.

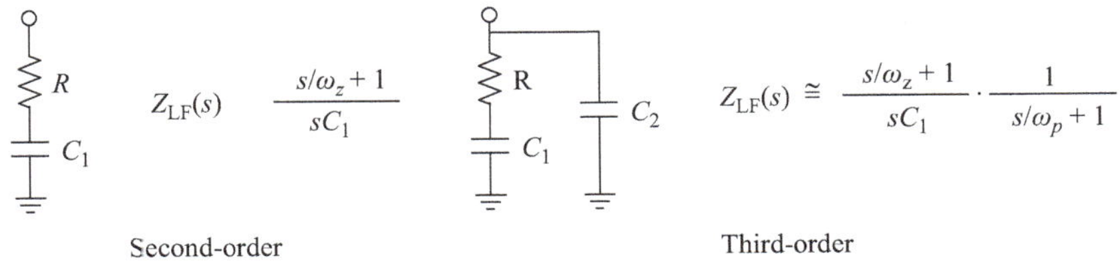

Figure 6.17 Loop filter impedance of the second-order loop filter and the third-order loop filter

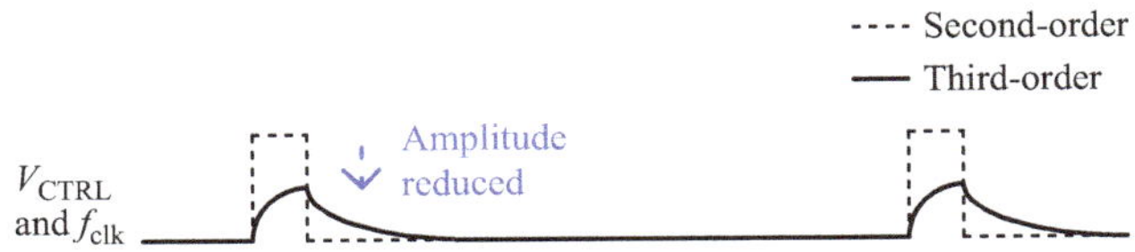

Figure 6.18 Reference spur reduction by C_2

Therefore, the C_2 is usually much smaller than C_1 not to degrade the phase margin. Then (6.32) can be simplified to

$$Z_{\mathrm{LF,3rd}}(s) \cong \frac{\frac{s}{\omega_z}+1}{sC_1}\cdot\frac{1}{\frac{s}{\omega_p}+1} = Z_{\mathrm{LF,2nd}}(s)\cdot\frac{1}{\frac{s}{\omega_p}+1} \tag{6.35}$$

where we can find that the only a pole is added over the impedance of the third-order loop filter. It means that the waveform of the V_{CTRL} of the third-order PLL is a low-pass filtered one from the second-order PLL's V_{CTRL}. Therefore, as shown in Figure 6.18, the third-order PLL is able to smoothen the periodic ripples in Figures 6.14–6.16, which cause the reference spur.

As observed in Figure 6.18, adding C_2 has a similar effect of reducing R as well as increasing C_1, which leads to degradation of the phase margin according to (6.17). Therefore, the value of C_2 should be carefully determined. Replacing the impedance of the second-order filter in (6.11) to (6.35), the open-loop transfer function of the third-order PLL becomes

$$T(s) = \frac{I_{\mathrm{CP}}}{2\pi}\left(\frac{\frac{s}{\omega_z}+1}{sC_1}\cdot\frac{1}{\frac{s}{\omega_p}+1}\right)\frac{2\pi K_{\mathrm{VCO}}}{s} = \frac{I_{\mathrm{CP}}K_{\mathrm{VCO}}}{s^2C_1}\cdot\frac{\frac{s}{\omega_z}+1}{\frac{s}{\omega_p}+1} \tag{6.36}$$

Substituting (6.36) into (6.10) leads to

$$\frac{I_{\mathrm{CP}}K_{\mathrm{VCO}}}{\omega_c^2(C_1+C_2)}\cdot\frac{\sqrt{\left(\frac{\omega_c}{\omega_z}\right)^2+1}}{\sqrt{\left(\frac{\omega_c}{\omega_p}\right)^2+1}} \cong N \tag{6.37}$$

Assuming that $\frac{\omega_c}{\omega_z} \gg 1$ and $\frac{\omega_c}{\omega_p} \ll 1$, ω_c of the third-order PLL is obtained approximately as

$$\omega_c = \frac{I_{CP} R K_{VCO}}{N\left(1 + \frac{C_2}{C_1}\right)} \tag{6.38}$$

which is scaled by $(1 + C_2/C_1)$ from that of the second-order PLL in (6.15). From (6.36), we can also calculate the phase margin of the third-order PLL as

$$\mathrm{PM} = \arctan\left(\frac{\omega_c}{\omega_z}\right) - \arctan\left(\frac{\omega_c}{\omega_p}\right) \tag{6.39}$$

Differentiating (6.39) leads to

$$\frac{d}{d\omega_c}\mathrm{PM} = \frac{1}{\omega_z} \cdot \frac{1}{1 + \left(\frac{\omega_c}{\omega_z}\right)^2} - \frac{1}{\omega_p} \cdot \frac{1}{1 + \left(\frac{\omega_c}{\omega_p}\right)^2} \tag{6.40}$$

By equating (6.40) to zero, we can derive the ω_c for maximizing the phase margin as

$$\omega_c = \sqrt{\omega_z \omega_p} \tag{6.41}$$

which is the geometric mean of ω_z and ω_p. We can also get an intuitive understanding of the phase margin of the third-order PLL from Figure 6.19, where the Bode plot of the open-loop magnitude response and the phase response are illustrated. Approximately, the phase starts increasing at $\omega_z/10$ and hits $+45°$ at ω_z, and eventually reaches $+90°$. The trend is similar for the ω_p, but the sign of the phase response is negative. Looking into the phase response of the third-order PLL,

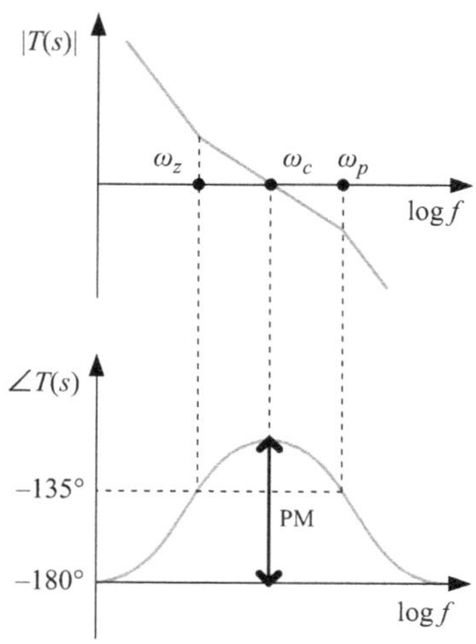

Figure 6.19 Bode plot of the third-order PLL open-loop transfer function

the phase response starts from $-180°$ because of two poles at DC. The phase increases as the frequency increases and becomes $-135°$ at the ω_z. The phase keeps increasing until the effect of the ω_p starts overtaking that of the ω_z, that is, at the middle of the ω_z and ω_p. Note that in logarithmic scale, the middle of the ω_z and ω_p equals the geometric mean of the ω_z and ω_p.

6.2.6 Bang-bang PLL

Due to its quantization nature, a bang-bang PLL exhibits a different aspect over a linear PLL under locked. Once achieving lock, ideally, a linear phase detector does not produce any output so there is no fluctuation on the output phase and frequency. On the other hand, however, a bang-bang phase detector always makes a quantized binary decision regardless of how low the phase difference is. As a result, the bang-bang phase detector produces alternate UP-DN sequences to nullify the average as shown in Figure 6.20(a). Such dithering between UP and DN states leads to a periodic fluctuation on the output phase whose amount is proportional to the

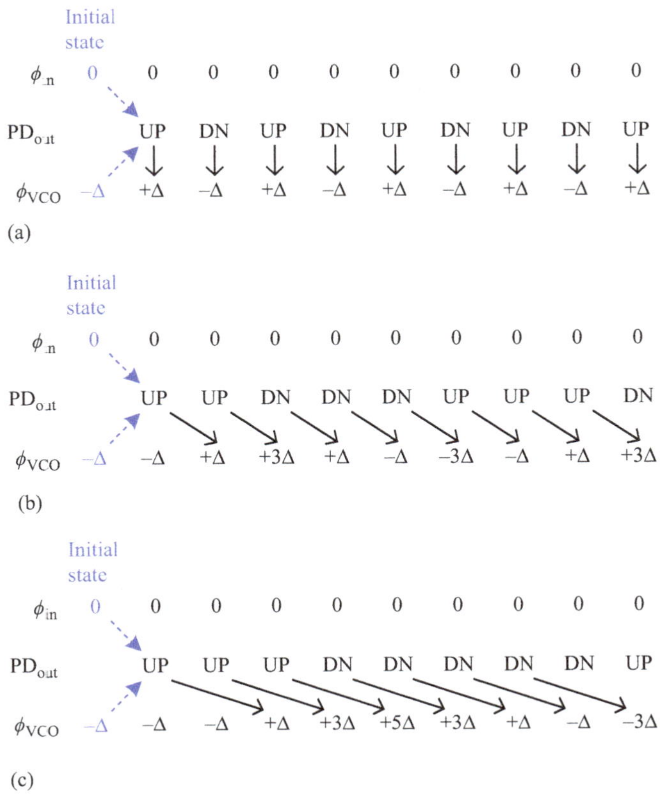

Figure 6.20 Bang-bang dithering and latency dependence (a) Latency = 0 UI (b) Latency = 1 UI (c) Latency = 2 UI

proportional gain of the loop filter and K_{VCO}. In addition, the bang-bang dithering is also a function of the loop latency [6,7]. For example, if the loop latency is 1 UI, the output phase is updated 1 UI after the phase detector makes the decision as shown in Figure 6.20(b). As a result, the phase detector produces two consecutive UPs if the initial phase difference is a small negative ($-\Delta$). After the consecutive UPs are updated, the phase error becomes $+3\Delta$ instead of $+\Delta$; so, for the next three cycles, the phase detector produces consecutive DNs, if we assume the phase step by one UP/DN is 2Δ. At the same manner, the latency of 2 UI results in five consecutive UPs/DNs and the maximum phase error of 5Δ. From this observation, we can find a general relation between the peak-to-peak dithering jitter and the latency as

$$J_{\text{p2p,dith}} = (1 + 2D) \cdot \Delta t \tag{6.42}$$

where D and Δt denote the loop latency and the phase step of dithering, respectively.

Another important characteristic of bang-bang PLL is that the loop gain is dependent on jitter. Even though we neglect the metastability of the bang-bang phase detector, jitter of either input or VCO clock causes a probability of false decision. As described in Section 5.2.2, the phase detector gain is defined as an average output across the phase difference. That is, if the phase error is less than the peak-to-peak jitter, there is a possibility of false decision so the output average cannot be 1 (or 0) but less than 1 (or higher than 0), as shown in Figure 6.21. With jitter distribution function $f(x)$, the output average of a bang-bang phase detector can be expressed as

$$E(\text{out}) = \int_{-\infty}^{\Delta\phi} f(x)dx \tag{6.43}$$

where $E(\text{out})$ is the expected value of the phase detector output, and we assume that the corresponding outputs for UP and DN are 1 and 0, respectively. Assuming a Gaussian distribution for the jitter which is given as

$$f(\Delta\phi) = \frac{1}{\sigma\sqrt{2\pi}} e^{-\frac{1}{2}\left(\frac{\Delta\phi}{\sigma}\right)^2} \tag{6.44}$$

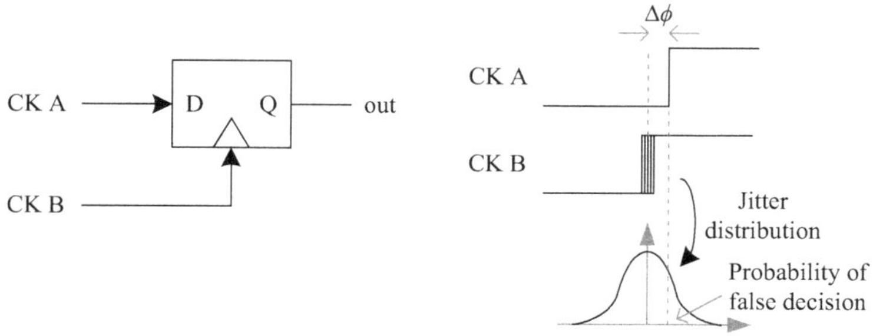

Figure 6.21 Bang-bang phase detector and false decision probability due to jitter

where σ is the standard deviation or the RMS jitter. Substituting (6.44), (6.43) becomes:

$$E(\text{out}) = \frac{1}{\sqrt{2\pi}} \int_{-\infty}^{\Delta\phi} e^{-\frac{x^2}{2}} dx \qquad (6.45)$$

which is a well-known cumulative distribution function (CDF) of the normal distribution, the Q function. Equation (6.45) is illustrated in Figure 6.22 for various jitter condition, where we can find that a kind of linearization happens to the gain curve, especially for the low phase error region (i.e. $-\sigma < \Delta\phi < \sigma$) [10–13]. For a very small phase error ($\Delta\phi \approx 0$), the linearized bang-bang gain is obtained by differentiating (6.45) as

$$\frac{d}{d\Delta\phi} E(\text{out})_{\Delta\phi=0} = f(\Delta\phi)_{\Delta\phi=0} = \frac{1}{\sigma\sqrt{2\pi}} \qquad (6.46)$$

Note that it is expected that a bang-bang PLL locks at this condition at the steady-state. Equation (6.46) shows the linearized gain is inversely proportional to RMS jitter. Here, the RMS jitter includes the jitter from the reference clock and the PLL clock so that it can be written to

$$\sigma = \sqrt{\sigma_{\text{ref}}^2 + \sigma_{\text{PLL}}^2 - 2\text{COV}_{\text{ref,PLL}}} \qquad (6.47)$$

where $\text{COV}_{\text{ref,PLL}}$ denotes the covariance of the reference jitter and the PLL jitter. Here we can find an important design challenge of a bang-bang PLL. As we have discussed in Section 6.2.3, the minimum jitter is achieved with the optimum loop bandwidth of a PLL that locates at the cross-point of the input jitter and the VCO jitter. However, in a bang-bang PLL, the loop bandwidth is a function of the jitters so that it makes it hard to optimize the loop bandwidth of a bang-bang PLL.

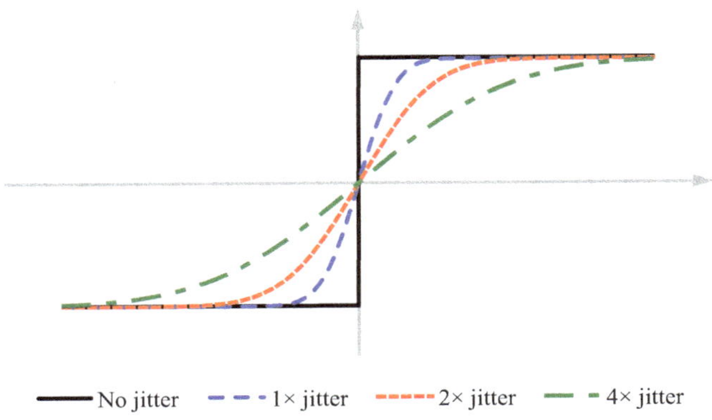

Figure 6.22 Linearization of bang-bang phase-detection gain by jitter

Moreover, increasing the loop gain also results in a higher dithering jitter, which makes the relation between the loop bandwidth and the PLL jitter more complex. To address this complexity, several background loop-bandwidth calibration techniques have been proposed in literatures [7–9] and readers can consult with if interested.

6.3 Supply noise-induced jitter

6.3.1 Impact of supply noise to PLL jitter

In addition to the noise from circuit elements, fluctuations in supply voltage also result in a considerable amount of jitter [14,15]. For example, the delay of a CMOS inverter is highly dependent on the supply voltage. Approximately, the delay can be obtained from the charge equation:

$$\frac{1}{2}\beta(V_{\mathrm{DD}} - V_{\mathrm{TH}})^{\alpha} \cdot D = \frac{1}{2}CV_{\mathrm{DD}} \tag{6.48}$$

where α ($1 < \alpha < 2$) and β are coefficients for MOSFET current expression. Equating (6.48) leads to

$$D = \frac{CV_{\mathrm{DD}}}{\beta(V_{\mathrm{DD}} - V_{\mathrm{TH}})^{\alpha}} \tag{6.49}$$

Supply voltage sensitivity of CMOS inverter is obtained by differentiating (6.49) to

$$\frac{dD}{dV_{\mathrm{DD}}} = \frac{C}{\beta(V_{\mathrm{DD}} - V_{\mathrm{TH}})^{\alpha}}\left(1 - \frac{\alpha V_{\mathrm{DD}}}{V_{\mathrm{DD}} - V_{\mathrm{TH}}}\right) = D\left(\frac{1}{V_{\mathrm{DD}}} - \frac{\alpha}{V_{\mathrm{DD}} - V_{\mathrm{TH}}}\right) \tag{6.50}$$

In practical cases, (6.50) is always nonzero implying that the supply noise is proportionally converted to jitter by modulating the delay. For example, assuming $V_{\mathrm{DD}} = 3\sim4V_{\mathrm{TH}}$ and $\alpha \approx 1.5$, (6.50) gives the sensitivity of 1% delay per 1% supply variation, which is known to a common rule of thumb for the delay sensitivity [16,18]. When there is a sinusoidal supply fluctuation $\Delta V \sin(2\pi f_n t)$, the amplitude of supply-induced jitter is

$$J_{\mathrm{VDD,buf}} = \Delta V \cdot \frac{dD}{dV_{\mathrm{DD}}} \tag{6.51}$$

At a similar manner, but more seriously, supply noise on a ring oscillator introduces jitter. From (6.49), the oscillation frequency of N-stage ring oscillator is written as

$$f_0 = \frac{1}{2ND} = \frac{1}{2N} \cdot \frac{\beta(V_{\mathrm{DD}} - V_{\mathrm{TH}})^{\alpha}}{CV_{\mathrm{DD}}} \tag{6.52}$$

Frequency sensitivity (K_{VDD}) is obtained from (6.52) as

$$K_{VDD} = \frac{df_0}{dV_{DD}} = \frac{\beta}{2NC} \cdot \frac{(V_{DD} - V_{TH})^{\alpha-1}}{V_{DD}} \left(\alpha - \frac{V_{DD} - V_{TH}}{V_{DD}} \right)$$
$$= f_0 \cdot \left(\frac{\alpha}{V_{DD} - V_{TH}} - \frac{1}{V_{DD}} \right) \tag{6.53}$$

Equations (6.52) and (6.53) are visualized in Figure 6.23 with respect to the V_{DD} for various α. Note that (6.50) also has the same shapes as Figure 6.23(b). In contrast to the former example, a sinusoidal supply noise is proportionally translated to the frequency error of

$$\Delta f_0 = \Delta V \cdot K_{VDD} \cdot \sin(2\pi f_n t). \tag{6.54}$$

The amplitude of the jitter is obtained by integrating the frequency error over a half period of the noise frequency:

$$J_{VDD,osc} = \frac{1}{f_0} \cdot \int_0^{\frac{1}{2f_n}} \Delta V \cdot K_{VDD} \cdot \sin(2\pi f_n t) dt = \Delta V \cdot \frac{1}{f_0} \cdot \frac{1}{\pi f_n} \cdot K_{VDD} \tag{6.55}$$

Comparing the amplitude of the supply-induced jitter of (6.51) and (6.55), we obtain:

$$\frac{J_{VDD,osc}}{J_{VDD,buf}} = \frac{1}{\pi f_n} \cdot \frac{1}{D} = \frac{2N}{\pi} \cdot \frac{f_0}{f_n} \tag{6.56}$$

In practice, $2N > \pi$ and $f_0 > f_n$, so we find that the ring oscillator is much sensitive to supply noise. Therefore, supply-noise rejection techniques are frequently used in low-jitter applications for ring oscillators and other sensitive circuits. In Section 6.3.2, we will study some well-known examples of supply-noise rejection circuits.

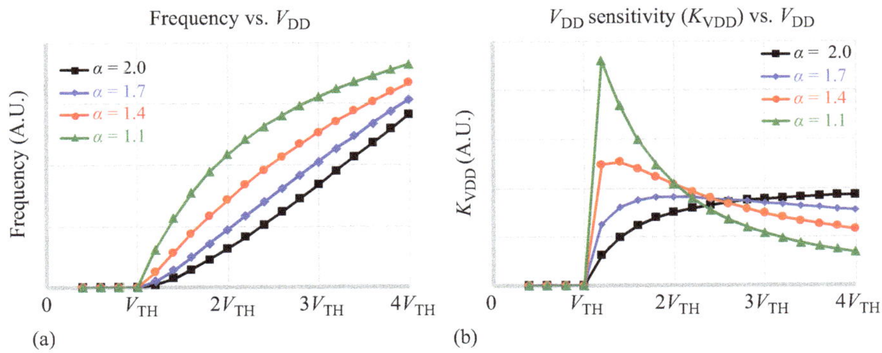

Figure 6.23 (a) Oscillation frequency and (b) supply sensitivity of ring oscillator

6.3.2 Supply-induced jitter reduction techniques

Supply-induced jitter reduction techniques can be classified into two subcategories: suppressing the supply noise itself or reducing the supply sensitivity of a circuit with a compensation technique. Low-dropout (LDO) regulator is a representative example of supply noise suppression techniques [17,18]. A simplified circuit diagram of an LDO regulator is shown in Figure 6.24(a). It is based on a negative feedback. The current consumed by the circuit load is provided through a pass transistor, but a negative feedback with an amplifier keeps the regulated supply voltage (V_{out}) equal to the reference voltage (V_{ref}) within the current range that the pass gate can support. A PMOS pass transistor is assumed for a low voltage drop since the supply sensitivity of the clocking circuit becomes greater with a lower V_{out} [18]. To analyze the LDO, we find two gain stages (A_1 and A_2) and two poles (ω_1 and ω_2), from the feedback amplifier and the pass transistor, respectively. The poles are formed by the output resistance of the gain stages (r_{o1} and r_{o2}) and the corresponding nodal capacitances (C_1 and C_2). In order to make the negative feedback stable, one of the poles must dominate the other. A linearized model of the LDO regulator is illustrated in Figure 6.24(b), where $H_1(s)$ and $H_2(s)$ are given as

$$H_1(s) = \frac{A_1}{1 + s/\omega_1} \tag{6.57}$$

$$H_2(s) = -\frac{A_2}{1 + s/\omega_2} \tag{6.58}$$

The transfer function of supply noise to the V_{out} is obtained as

$$H_{reg}(s) = \frac{\Delta V_{out}}{\Delta V_{DD}} = \frac{-H_2(s)}{1 - H_1(s)H_2(s)} = \frac{\frac{A_2}{1+s/\omega_2}}{1 + \frac{A_1}{1+s/\omega_1} \cdot \frac{A_2}{1+s/\omega_2}} \tag{6.59}$$

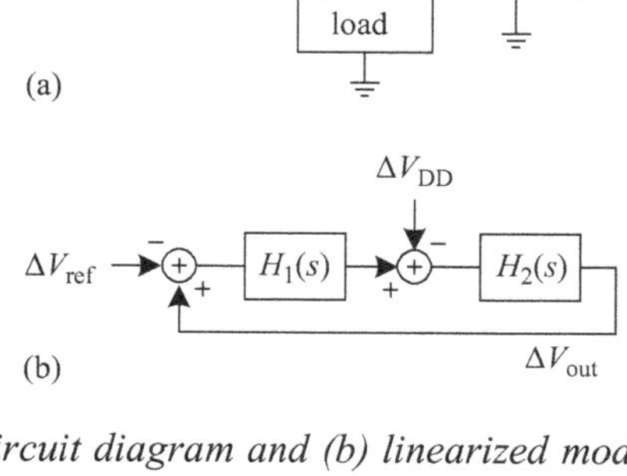

Figure 6.24 (a) Circuit diagram and (b) linearized model of LDO regulator

The magnitude response of (6.59) is visualized in Figure 6.25, where k denotes the ratio of ω_2/ω_1. At low frequency, (6.59) is simplified to

$$H_{\text{reg}}(s) \approx \frac{A_2}{1 + A_1 A_2} \tag{6.60}$$

which is further simplified to $1/A_1$ assuming enough loop gain, where we find that the low-frequency supply noise is corrected by the negative feedback by a factor of the amplifier gain. On the other hand, at very high frequency, it is simplified as

$$H_{\text{reg}}(s) \approx \frac{A_2}{1 + s/\omega_2} \tag{6.61}$$

which implies the high-frequency noise is filtered out by the finite bandwidth of the second gain stage while the feedback loop is not able to react to such high-frequency noise. On the other hand, for intermediate frequency region, two different cases of $\omega_2 \ll \omega_1$ and $\omega_2 \gg \omega_1$ are considered based on stability requirement. For the first case, (6.59) is simplified as

$$H_{\text{reg}}(s) \approx \frac{\frac{A_2}{1+s/\omega_2}}{1 + \frac{A_1 A_2}{1+s/\omega_2}} = \frac{A_2}{1 + A_1 A_2 + s/\omega_2} \approx \frac{1}{A_1(1 + s/A_1 A_2 \omega_2)} \tag{6.62}$$

The implication of (6.62) is that the transfer function is approximated to a first-order low-pass filter whose cut-off frequency is $A_1 A_2 \omega_2$ when we have the dominant pole at the output of the regulator. The plot for $k = 0.01$ in Figure 6.25 corresponds to this case. On the other hand, for the second case where the dominant pole is at the output of the amplifier, we simplify (6.59) to

$$H_{\text{reg}}(s) \approx \frac{\frac{A_2}{1+s/\omega_2}}{1 + \frac{A_1 A_2}{1+s/\omega_1}} = \frac{\frac{A_2(1+s/\omega_1)}{1+s/\omega_2}}{1 + A_1 A_2 + s/\omega_1} \approx \frac{(1+s/\omega_1)}{A_1\left(1 + \frac{s}{A_1 A_2 \omega_1}\right)(1+s/\omega_2)} \tag{6.63}$$

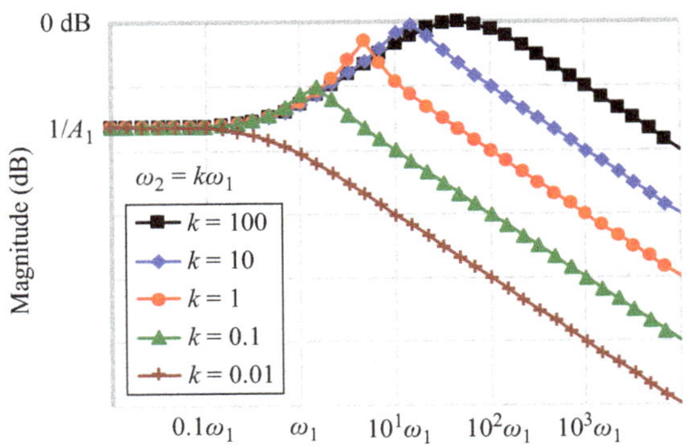

Figure 6.25 Power supply noise rejection of LDO regulator

where we find one zero at ω_1 and two poles at ω_2 and $A_1A_2\omega_1$. As a result, the transfer gain starts increasing once the frequency is higher than ω_1 but becomes flat after hitting the first pole. At high-frequency after the second pole, (6.63) decreases at -20 dB/dec slope. This case corresponds to the $k = 100$ in Figure 6.25. We can observe that the dominant pole at the regulator output exhibits a better supply noise rejection; however, it is much expensive. Since the pass transistor draws a large current, it has a large (W/L) so r_{o2} is generally much smaller than r_{o1}. Therefore, a huge capacitance is required for C_2 to make ω_2 be the dominant one. Note that the gate capacitance of such large pass transistor introduces a large intrinsic capacitance for ω_1. Usually such high capacitance is not affordable with on-chip capacitor. On the other hand, having dominant pole at the output of the amplifier can be enabled much efficiently. At first, the r_{o1} is larger than r_{o2} so that a relatively smaller C_1 is enough for stability requirements. And second, stability-enhancing techniques such as Miller capacitance or pole-zero compensation which utilize the pass transistor stage further reduce the capacitance required to meet stability requirements. To summarize, LDO regulators have a trade-off between supply noise rejection and cost, so careful choice and design are required based on full considerations on application and specification. Designers should also consider that the voltage drop of an LDO results in a power loss across the pass transistor and a higher supply sensitivity as expected from (6.50) and (6.53).

As an alternate of suppressing the supply noise itself, we can try to make buffer delay insensitive to the supply variation. The concept of supply-insensitive delay cell is shown in Figure 6.26(a). Since the intrinsic delay has a negative sensitivity to the supply voltage variation, zero sensitivity is achieved by introducing an additional delay with a positive sensitivity. Figure 6.26(b)–(d) shows three examples of supply-sensitivity compensated delay cells [19,21]. The first example shown in Figure 6.26(b) is based on capacitive compensation. A series RC load is added at the output of a CMOS inverter, where the resistor is implemented with a PMOS transistor (P2) whose gate bias (V_B) is a reference to the ground. The effective capacitive loading of the RC load depends on the P2 resistance. For examples of two extreme cases, the capacitance equals C if the resistance is zero, but it becomes zero if the resistance goes to infinite. When the V_{DD} raises, the gate-overdrive voltage of P2 increases so that the resistance becomes lower. As a result, the effective capacitance increases hence introducing a positive sensitivity to the supply voltage. The sensitivity is a function of V_B, P2 sizing, and C. The second example shown in Figure 6.26(c) relies on a current-based compensation instead of the capacitance-based one. Rather than using a CMOS inverter, the main NMOS (N1) receives the input signal but the PMOS (P1) is biased to V_{B1}, which references V_{DD}. There is another pull-down path with N2 whose gate bias (V_{B2}) comes from a source of N3, which is biased with a current source. Since gate and drain of the N3 are tied to V_{DD}, it operates as a source-follower so that V_{DD} fluctuation directly appears to V_{B2}. As a result, N2 flows more current with positive ΔV_{DD}, and therefore it slows down the pull-up transition by the P1. A simpler version shown in Figure 6.26(d), where the sink path is removed, also provides a compensated supply sensitivity. Because the PMOS bias references V_{DD}, the PMOS pull-up current is

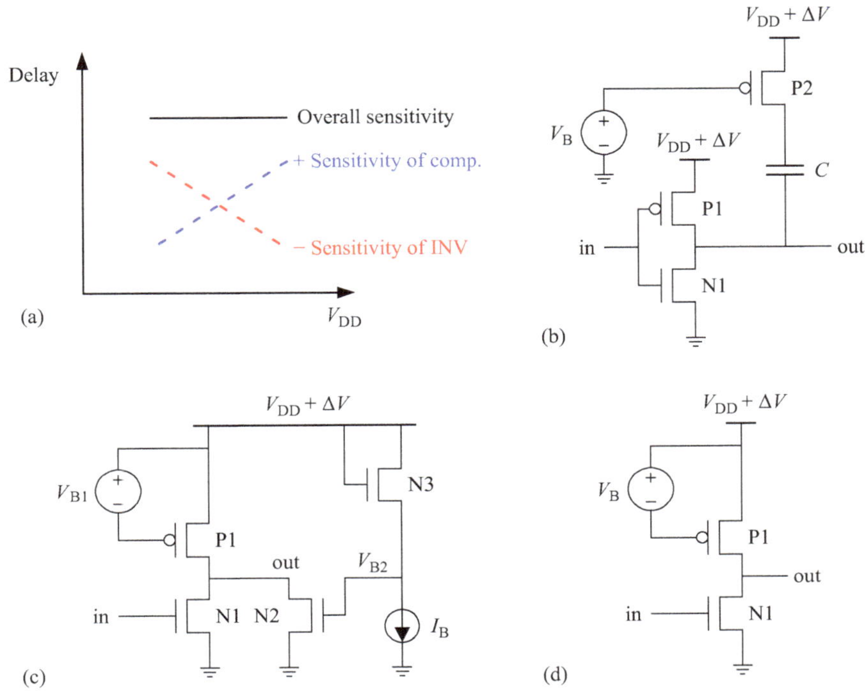

Figure 6.26 Supply sensitivity compensation techniques: (a) basic concept, (b) circuits proposed in [19], (c) in [20], and (d) in [21]

relatively constant regardless of ΔV_{DD} assuming operating in the saturation region. At the same time, the voltage swing increases by ΔV_{DD}. As a result, the pull-up delay variation due to ΔV_{DD} can be written:

$$\Delta D_{up} = \frac{C \cdot \Delta V_{DD}}{I} \tag{6.64}$$

where we can find a positive sensitivity.

Appendix A: Analytic expression of the reference spur

In this appendix, the analytic expression of the reference spur due to the loop filter leakage is derived as an example [22]. If there is nonzero leakage current through the loop filter capacitor, the difference between UP and DN pulses (ΔT) can be calculated as

$$\Delta T = \frac{I_{leak}}{I_{CP,0}} T_{ref} \tag{A.1}$$

where T_{ref} is the period of the reference clock, I_{leak} is the leakage current of the loop filter, and the $I_{\text{CP},0}$ is the amplitude of the current of the CP. Using the Fourier series, $I_{\text{CP}}(t)$ is expressed as

$$I_{\text{CP}}(t) = \sum_{n \neq 0} \frac{2I_{\text{CP},0}}{n\pi} \cdot \sin\left(n\pi \frac{I_{\text{leak}}}{I_{\text{CP},0}}\right) \cdot \cos(2n\pi f_{\text{ref}} t) \tag{A.2}$$

Let us assume that we can neglect the harmonic components for simplicity. Then, the fundamental frequency component of the CP current is achieved by neglecting harmonic tones in (A.2). Therefore, from (A.2), V_{ctrl} becomes:

$$V_{\text{ctrl}}(t) = V_0 + \frac{2I_{\text{CP},0}}{\pi} \cdot \sin\left(\pi \frac{I_{\text{leak}}}{I_{\text{CP},0}}\right) \cdot \cos(2\pi f_{\text{ref}} t) \cdot Z_{\text{LF}}(s) \tag{A.3}$$

where V_0 is the DC component of V_{ctrl}. Since the output frequency of PLL, f_{clk}, is proportional to V_{ctrl}, the reference spur can be derived using the frequency modulation theory [23]. Assuming that the modulation index β is sufficiently small, the reference spur amplitude is approximately half of the modulation index ($J_1(\beta) \cong \beta/2$). Therefore, the reference spur amplitude is given as

$$\text{Spur} = 20\log\left(\frac{I_{\text{CP},0}}{\pi} \cdot \sin\left(\pi \frac{I_{\text{leak}}}{I_{\text{CP},0}}\right) \cdot |Z_{\text{LF}}(f_{\text{ref}})| \cdot \frac{K_{\text{VCO}}}{f_{\text{ref}}}\right) \tag{A.4}$$

where K_{VCO} is the VCO gain in Hz/V. Using the open-loop gain of the PLL in (6.11), (A.4) is simplified to

$$\text{Spur} = 20\log\left(\frac{2}{\pi} T(f_{\text{ref}}) \cdot \sin\left(\pi \frac{I_{\text{leak}}}{I_{\text{CP},0}}\right)\right) \tag{A.5}$$

Assuming $I_{\text{leak}} \ll I_{\text{CP},0}$,

$$\text{Spur} = 20\log\left(2 \cdot T\left(f_{ref}\right) \cdot \frac{I_{\text{leak}}}{I_{\text{CP},0}}\right) \tag{A.6}$$

where we find that the reference spur is proportional to the I_{leak}, and a larger pump current is beneficial to suppress the reference spur.

Appendix B: Why do we use PLL rather than FLL for frequency generation?

As discussed in the previous chapters, the major purpose of clock generators is to generate a precise frequency, and the phase alignment techniques are used to realize that. If so, one may ask why we do not use a frequency-locked loop (FLL) instead of a PLL. In fact, an ideal FLL is able to provide a clean clock reference. Here, the ideal FLL means that the output of the FLL provides a pure single tone, whose frequency matches perfectly with the reference frequency. However, practical circuit implementations always introduce many sources of nonideality. For example, as we have discussed in Section 6.2.4, various mismatches are introduced

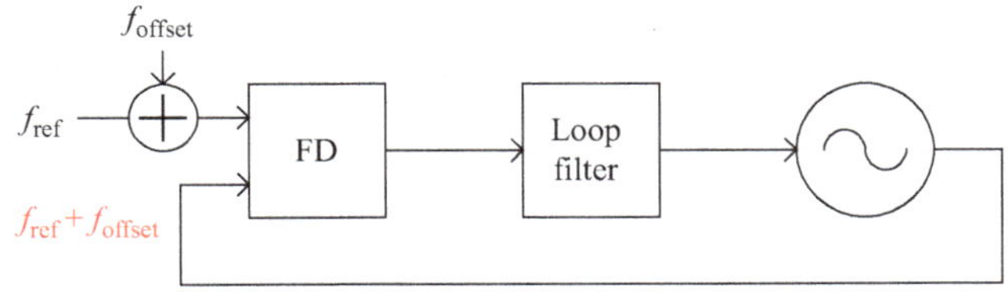

Figure B.1 Frequency-locked loop with frequency detector offset

from the practical implementation of phase detectors or charge-pumps. However, in a PLL, we studied that such mismatches introduce a static phase offset and a reference spur, but the average frequency perfectly locks to the input reference. On the other hand, any mismatch in FLL results in a frequency offset, rather than a phase offset. For example, if there is a static offset between the two inputs of a frequency detector, it is translated to an input-referred frequency offset, as shown in Figure B.1. That is, considering the nonidealities of practical circuit implementations, a PLL is able to limit those nonideal effects to the phase offset, whereas they result in the frequency offset when the FLL is used, which is not acceptable in most of the applications.

References

[1] Fischette D. First time, every time-practical tips for phase-locked loop design. *IEEE Distinguished Lecturer Ser. PLL Tutorial*. 2009.

[2] Gardner F. Charge-pump phase-lock loops. *IEEE Transactions on Communications*. 1980;28(11):1849–1858.

[3] Gao X, Klumperink EA, Socci G, Bohsali M, Nauta B. Spur reduction techniques for phase-locked loops exploiting a sub-sampling phase detector. *IEEE Journal of Solid-State Circuits*. 2010;45(9):1809–1821.

[4] Wang KJ, Swaminathan A, Galton I. Spurious tone suppression techniques applied to a wide-bandwidth 2.4 GHz fractional-N PLL. *IEEE Journal of Solid-State Circuits*. 2008;43(12):2787–2797.

[5] Lo SH, Buchanan DA, Taur Y, Wang W. Quantum-mechanical modeling of electron tunneling current from the inversion layer of ultra-thin-oxide nMOSFET's. *IEEE Electron Device Letters*. 1997;18(5):209–211.

[6] Da Dalt N. A design-oriented study of the nonlinear dynamics of digital bang-bang PLLs. *IEEE Transactions on Circuits and Systems I: Regular Papers*. 2005;52(1):21–31.

[7] Jang S, Kim S, Chu SH, Jeong GS, Kim Y, Jeong DK. An optimum loop gain tracking all-digital PLL using autocorrelation of bang–bang phase-frequency detection. *IEEE Transactions on Circuits and Systems II: Express Briefs*. 2015;62(9):836–840.

[8] Liang J, Sheikholeslami A, Tamura H, Ogata Y, Yamaguchi H. Loop gain adaptation for optimum jitter tolerance in digital CDRs. *IEEE Journal of Solid-State Circuits*. 2018;53(9):2696–2708.

[9] Marucci G, Levantino S, Maffezzoni P, Samori C. Exploiting stochastic resonance to enhance the performance of digital bang-bang PLLs. *IEEE Transactions on Circuits and Systems II: Express Briefs.* 2013;60(10): 632–636.

[10] Lee J, Kundert KS, Razavi B. Analysis and modeling of bang-bang clock and data recovery circuits. *IEEE Journal of Solid-State Circuits.* 2004;39(9): 1571–1580.

[11] Yoo BJ, Bae WR, Han J, Kim J, Jeong DK. Linearization technique for binary phase detectors in a collaborative timing recovery circuit. *IEEE Transactions on Very Large Scale Integration (VLSI) Systems.* 2013;22(6):1226–1237.

[12] Park MJ, Kim J. Pseudo-linear analysis of bang-bang controlled timing circuits. *IEEE Transactions on Circuits and Systems I: Regular Papers.* 2012;60(6):1381–1394.

[13] Da Dalt N. Linearized analysis of a digital bang-bang PLL and its validity limits applied to jitter transfer and jitter generation. *IEEE Transactions on Circuits and Systems I: Regular Papers.* 2008;55(11):3663–3675.

[14] Herzel F, Razavi B. A study of oscillator jitter due to supply and substrate noise. *IEEE Transactions on Circuits and Systems II: Analog and Digital Signal Processing.* 1999;46(1):56–62.

[15] Heydari P, Pedram M. Analysis of jitter due to power-supply noise in phase-locked loops. In *Proceedings of the IEEE 2000 Custom Integrated Circuits Conference (Cat. No. 00CH37044)*. Orlando, FL: IEEE; 2000 (pp. 443–446).

[16] Sidiropoulos S, Liu D, Kim J, Wei G, Horowitz M. Adaptive bandwidth DLLs and PLLs using regulated supply CMOS buffers. In *2000 Symposium on VLSI Circuits. Digest of Technical Papers (Cat. No. 00CH37103).* Honolulu, HI: IEEE; 2000 (pp. 124–127).

[17] Alon E, Kim J, Pamarti S, Chang K, Horowitz M. Replica compensated linear regulators for supply-regulated phase-locked loops. *IEEE Journal of Solid-State Circuits.* 2006;41(2):413–424.

[18] Casper B, O'Mahony F. Clocking analysis, implementation and measurement techniques for high-speed data links: A tutorial. *IEEE Transactions on Circuits and Systems I: Regular Papers.* 2009;56(1):17–39.

[19] Mansuri M, Yang CK. A low-power adaptive bandwidth PLL and clock buffer with supply-noise compensation. *IEEE Journal of Solid-State Circuits.* 2003;38(11):1804–1812.

[20] Wu T, Mayaram K, Moon UK. An on-chip calibration technique for reducing supply voltage sensitivity in ring oscillators. *IEEE Journal of Solid-State Circuits.* 2007;42(4):775–783.

[21] Hsieh PH, Maxey J, Yang CK. Minimizing the supply sensitivity of a CMOS ring oscillator through jointly biasing the supply and control voltages. *IEEE Journal of Solid-State Circuits.* 2009;44(9):2488–2495.

[22] Ko HG, Bae W, Jeong GS, Jeong DK. Reference spur reduction techniques for a phase-locked loop. *IEEE Access.* 2019;7:38035–38043.

[23] Thomas TG, Sekhar SC. *Communication Theory.* New York: Tata McGraw-Hill; 2006.

Chapter 7
DLL loop dynamics and jitter

7.1 DLL basics

In this chapter, loop dynamics of DLLs are derived and compared with those of PLL. In general, DLLs are classified into type-I DLL and type-II DLL, according to the number of input clocks. Figure 7.1 shows simplified block diagrams of the type-I and type-II DLLs. In type-I DLL, there is only one input which is fed to the VCDL, and the phase detector compares the input and the output of the VCDL. The applications of type-I DLL include multiphase clock generation and zero-delay buffer. On the other hand, there are two inputs in the type-II DLL. Whereas the same clock signal is fed to both VCDL and phase detector in the type-I DLL, one of the type-II DLL input clocks is served only as the reference signal of the VCDL but the other is used to provide a reference phase that the VCDL output clock is driven to be aligned. That is, the VCDL delays only one of the inputs, and the phase detector compares the output of the VCDL and the other input. Main application of type-II DLL is clock and data recovery circuit in mesochronous clocking receivers.

In PLLs, phase alignment relies on a VCO which adjusts its output frequency according to the input voltage. As a result, a pole at zero frequency (0 Hz) is introduced in the phase-domain transfer function of a VCO, because the integration of frequency is phase. Therefore, there are at least two poles in PLL loop, and an inherent stability issue exists in PLLs. As a result, to guarantee sufficient stability, a zero should be added in the PLL loop filter. On the other hand, DLLs are unconditionally stable system because VCDL does not introduce a pole. The comparison of the VCO and the VCDL is illustrated in Figure 7.2. Therefore, an integrator is sufficient for the loop filter of a DLL. Since a charge-pump followed by a capacitor acts a perfect integrator, a charge-pump-based DLL is good to show general characteristics of DLL, so that it will be used to derive the loop dynamics in the remainder of this chapter.

7.2 DLL jitter

7.2.1 Input jitter transfer

Phase transfer function of a clocking circuit is very important because it reflects the jitter transfer or the filtering characteristic. In this chapter, transfer functions from

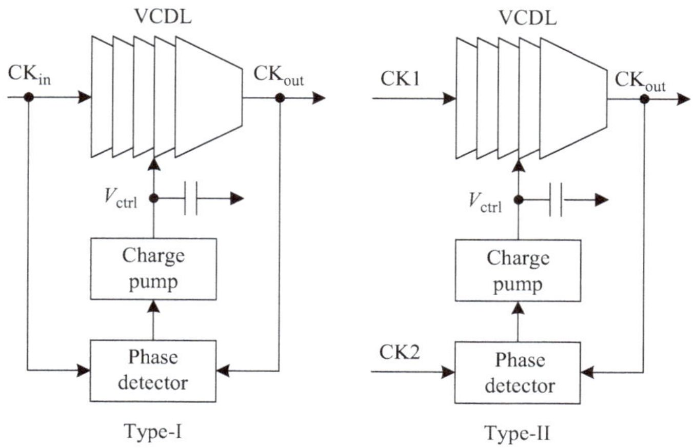

Figure 7.1 Block diagrams of type-I and type-II DLLs

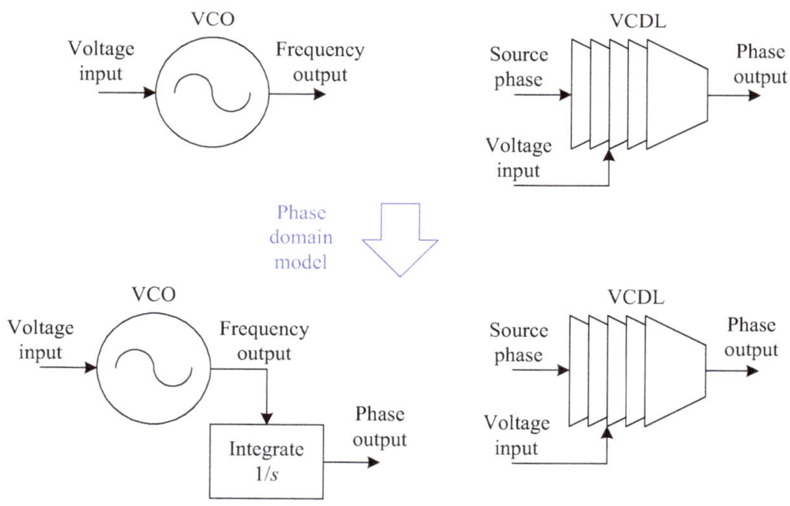

Figure 7.2 Comparison of VCO and VCDL

various jitter sources of DLL will be derived. Figure 7.3 shows the phase domain model and the derivation process of the input-to-output phase transfer function of the type-I DLL, which includes three steps. In the phase domain model shown in the first diagram, the gain of the phase detector and charge-pump (PD/CP) are merged for simplicity. Distinguished to the VCO where there is only one input

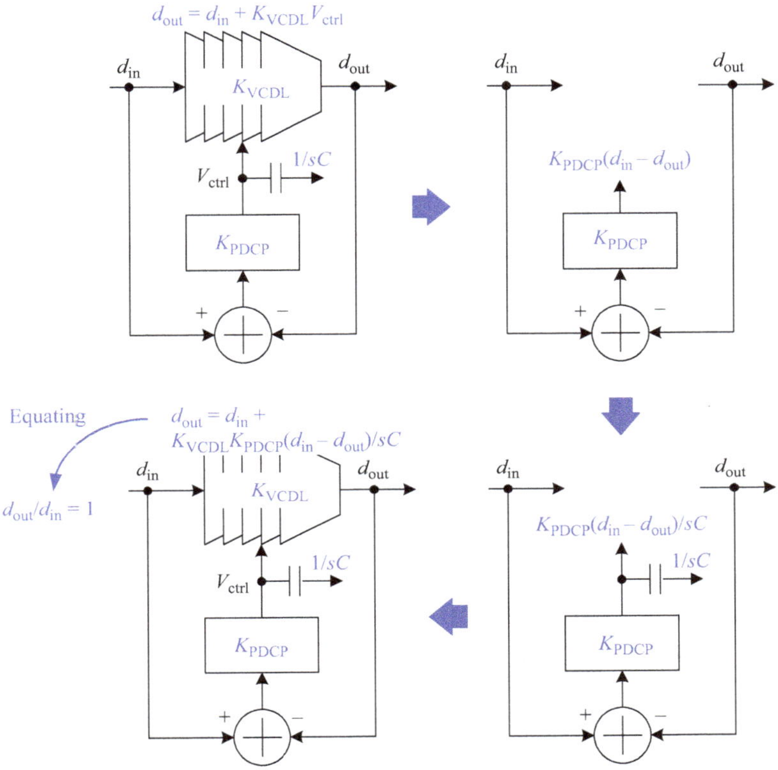

Figure 7.3 Derivation of input-to-output phase transfer function of type-I DLL

(V_{ctrl}), the VCDL has two inputs, the input delay (d_{in}) and the control voltage (V_{ctrl}). The output delay (d_{out}) of the VCDL is given as

$$d_{out} = d_{in} + K_{VCDL}V_{ctrl} \tag{7.1}$$

where K_{VCDL} is the linearized transfer gain of the VCDL (s/V). Note that V_{ctrl} sets the output delay relative to the input delay, whereas the VCO does not have any reference input phase or frequency. This aspect of VCDL will make a big difference over the PLL. In the first step of the derivation (see the top-right diagram of Figure 7.3), the input delay and output delay are subtracted and multiplied by the gain of the phase detector and charge pump. After that, the charge pump current is integrated by the loop filter capacitor. This integrated voltage is fed back to the VCDL, so that (7.1) can be rewritten as

$$d_{out} = d_{in} + K_{PDCP}K_{VCDL}(d_{in} - d_{out})/sC \tag{7.2}$$

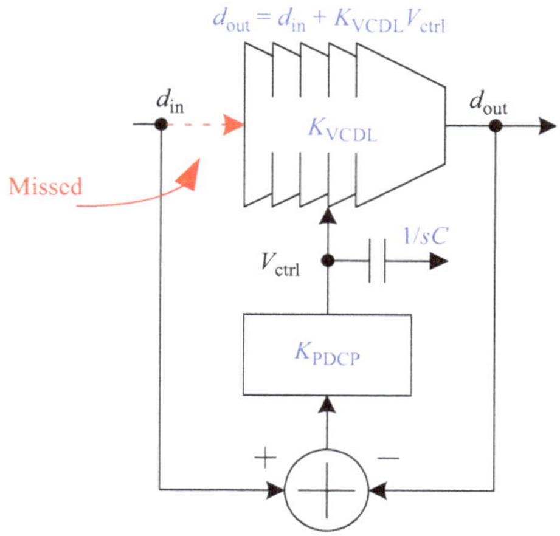

Figure 7.4 Missing path misleading DLL transfer function

By solving (7.2), the input-jitter transfer function of the type-I DLL is obtained as

$$H(s) = \frac{d_{out}}{d_{in}} = \frac{1 + K_{PDCP}K_{VCDL}/sC}{1 + K_{PDCP}K_{VCDL}/sC} = 1 \tag{7.3}$$

The transfer function is unity regardless of frequency, that is, the DLL is an all-pass filter [1]. This is a unique feature of the DLL; however, some literatures mislead that DLL is a low-pass filter in the phase domain analysis because they miss out the source clock path highlighted in Figure 7.4. It results in missing d_{in} in (7.1) and (7.2) and leads to a totally wrong transfer function, because a VCDL can never generate output without being fed an input signal, unlike the VCO. Again, a VCDL adjusts the relative phase of the output, referenced to the phase of the input. Neglecting the highlighted path results in omission of the reference phase, and therefore it cannot reflect the operation of the VCDL at all.

In PLLs, we studied that loop latency degrades the stability of the feedback loop. A huge latency can also introduce a stability issue to DLLs; however, its effect may not be as critical as that in PLLs, since DLLs are inherently much stable than PLLs. Instead, the latency effect on the jitter transfer is generally more focused when we deal with DLL design. Figure 7.5 shows a more realistic model of type-I DLL which includes the latency. Because the VCDL delays the input clock by a certain value, a latency element is included in the feedback path. With the latency element, e^{-sT}, the closed-loop transfer function of the DLL becomes:

$$H(s) = \frac{s + K_{PDCP}K_{VCDL}/C}{s + e^{-sT}K_{PDCP}K_{VCDL}/C} \tag{7.4}$$

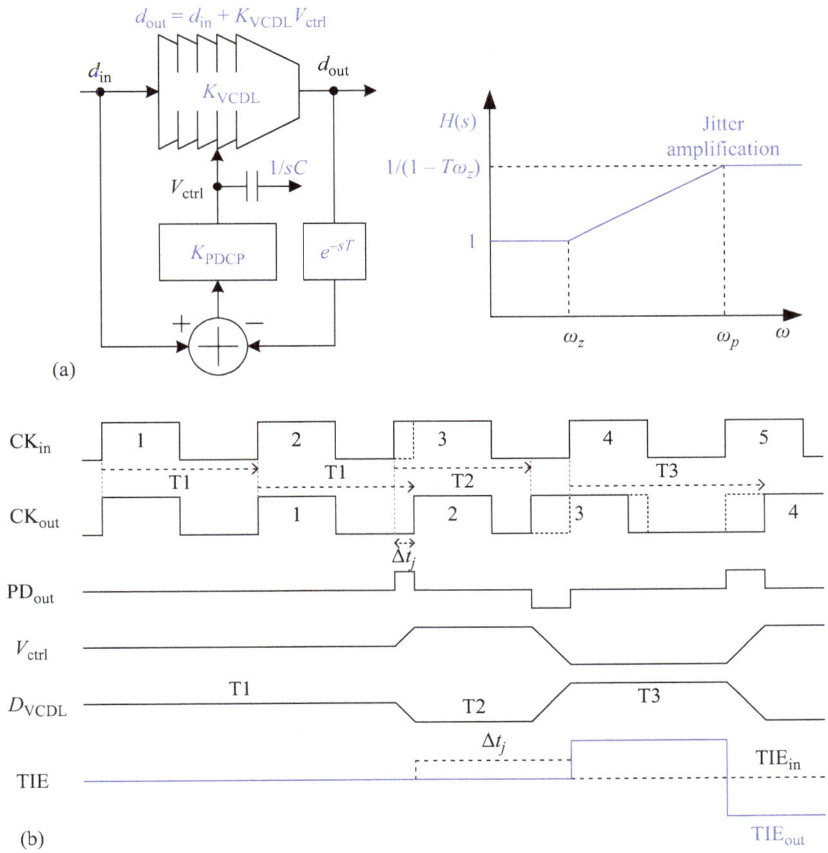

Figure 7.5 (a) DLL model and transfer function including latency element and (b) time domain interpretation of the jitter amplification

Assuming a small value of sT, the latency e^{-sT} can be replaced by $1 - sT$. Then (7.4) can be rewritten as

$$H(s) = \frac{s + K_{\mathrm{PDCP}}K_{\mathrm{VCDL}}/C}{s(1 - TK_{\mathrm{PDCP}}K_{\mathrm{VCDL}}/C) + K_{\mathrm{PDCP}}K_{\mathrm{VCDL}}/C} \tag{7.5}$$

The locations of the zero and pole are obtained as

$$\omega_z = K_{\mathrm{PDCP}}K_{\mathrm{VCDL}}/C \tag{7.6}$$

$$\omega_p = \frac{K_{\mathrm{PDCP}}K_{\mathrm{VCDL}}/C}{1 - TK_{\mathrm{PDCP}}K_{\mathrm{VCDL}}/C} = \frac{\omega_z}{1 - T\omega_z} \tag{7.7}$$

The Bode plot of (7.5) is illustrated in Figure 7.5(a), which shows that the transfer function is no longer flat. Because the denominator of (7.7) is always smaller than unity, the zero occurs at a lower frequency than the pole. As a result, the gain becomes larger than unity in the high-frequency region, which is referred to the jitter amplification, or the jitter peaking [2]. The amount of the jitter amplification equals $1/(1 - T\omega_z)$, which is obtained from the ratio of the pole frequency over the zero frequency. Therefore, the jitter amplification can be mitigated by lowering the open-loop gain, for example, reducing the charge pump current or increasing the loop filter capacitance. Note that reducing K_{VCDL} is generally not a viable option, because K_{VCDL} is set by the required VCDL tuning range and the available voltage headroom of the V_{ctrl}. More detailed explanation of VCDL design considerations will be covered in Section 7.4.

Time-domain representation of the jitter amplification shown in Figure 7.5(b) gives an intuitive way to understand the jitter transfer of the type-I DLL. A low-frequency jitter component of d_{in} and d_{out} is fully correlated so that the phase detector does not generate output at the steady-state. On the other hand, the correlation is reduced when the jitter frequency increases, due to the delay of the VCDL. For example, let us assume that the momentary jitter (Δt_j) makes d_{in} be smaller than the ideal value (see the third rising edge of the $\mathrm{CK_{in}}$ in Figure 7.5(b)). At the same time, however, d_{out} has not become smaller yet because there is latency (T) until the jitter arrives at the d_{out}. Therefore, the phase detector tries to decrease the VCDL delay because it thinks that d_{out} is larger than d_{in}, even though it is not a matter of the delay. The DLL loop gain sets the amount of the decreased delay. In the next cycle, the jitter finally passes through the VCDL hence makes d_{out} be even smaller than d_{in}, in addition to the decreased delay of the VCDL. As a result, the d_{out} experiences a larger jitter than the input jitter (Δt_j), and the additive amount is proportional to the loop gain.

Let us derive the input-jitter transfer function of the type-II DLL. Note that there are two input-to-output transfer functions because the type-II DLL has two inputs. The transfer function from CK1, which is the input clock of the VCDL, is derived in Figure 7.6, where d_1, d_2, and d_{out} denote the delay of CK1, CK2, and $\mathrm{CK_{out}}$, respectively. While deriving d_1 to $\mathrm{d_{out}}$ transfer, CK2 is ignored so that d_2 is assumed to be zero. The flow is of course very similar to the derivation for the type-I DLL. Since d_2 is zero, d_{out} is directly converted to the current through the phase detector and charge-pump, and then is integrated by the loop filter capacitor. Substituting the integrated voltage into V_{ctrl} of (7.1), we can obtain:

$$d_{\mathrm{out}} = d_1 - K_{\mathrm{PDCP}}K_{\mathrm{VCDL}}d_{\mathrm{out}}/sC \tag{7.8}$$

which leads to

$$H(s) = \frac{d_{\mathrm{out}}}{d_1} = \frac{s}{s + K_{\mathrm{PDCP}}K_{\mathrm{VCDL}}/C} \tag{7.9}$$

Contrary to the case of type-I DLL, a high-pass transfer is obtained. Intuitively, the feedback loop is able to track the low-frequency jitter component (low-frequency means lower than the cut-off frequency), that is the jitter can be fully canceled by the loop. For example, when a phase offset is introduced to d_1, the loop

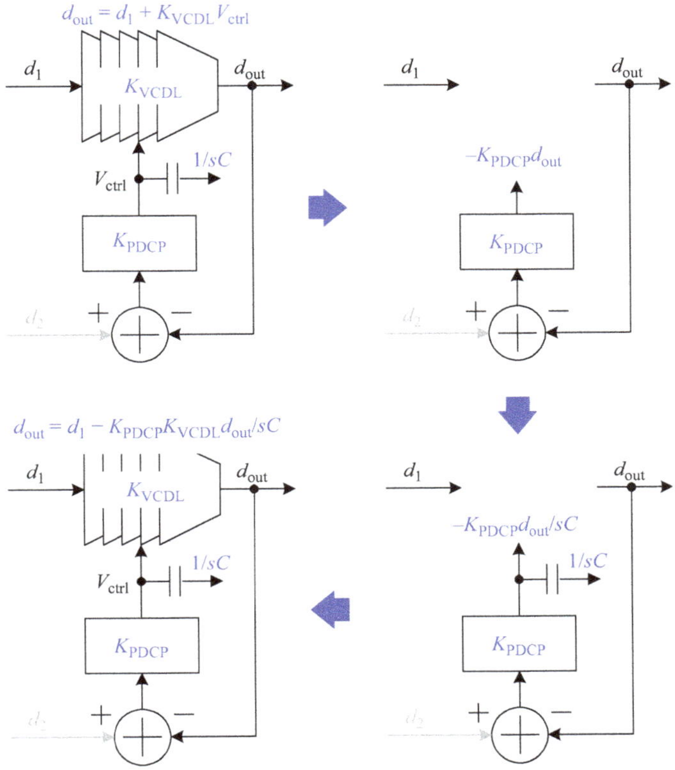

Figure 7.6 Derivation of CK1-to-output transfer function of type-II DLL

gradually increases (or decreases) the control voltage and eventually tracks the offset. Note that we can regard an offset as a kind of very slow noise. In contrast, the high-frequency jitter is not tracked by the loop so that the control voltage stays at a constant, which means the input jitter is directly reflected to the output through the fixed VCDL delay.

From Figure 7.7, on the other hand, the d_2 to d_{out} transfer function can be derived as

$$H(s) = \frac{d_{out}}{d_2} = \frac{K_{PDCP}K_{VCDL}/C}{s + K_{PDCP}K_{VCDL}/C} \tag{7.10}$$

That is, the jitter of CK2 is low-pass filtered because it passes the entire low-pass loop to reach the output whereas CK1 passes only the VCDL. Therefore, the jitter of CK2 is tracked by the loop, whereas the jitter of CK1 is canceled by the loop (and the loop is a low-pass filter). If the d_1 and d_2 are fully correlated, they can be regarded as the same input in terms of the jitter transfer. As a result, the overall transfer function can be simply obtained by summing (7.9) and (7.10), which leads

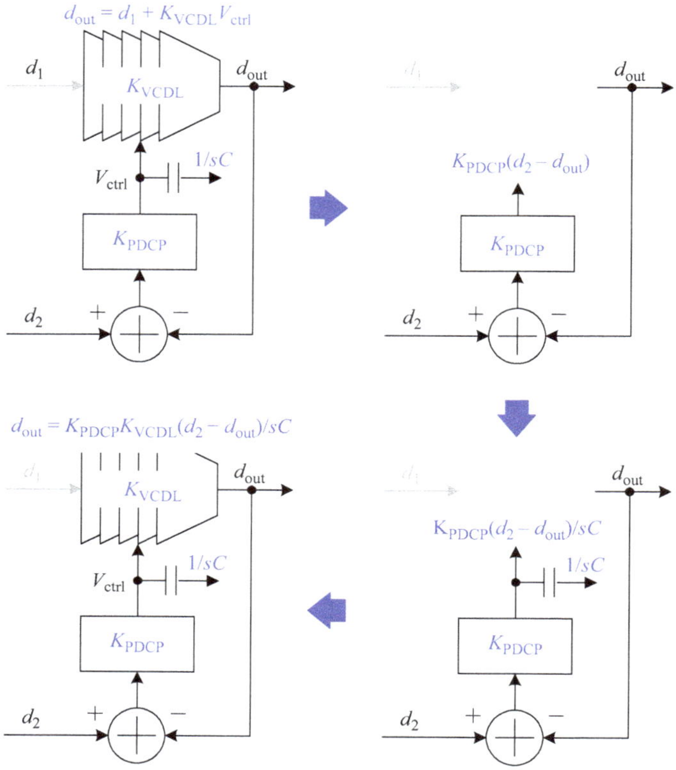

Figure 7.7 Derivation of CK2-to-output transfer function of type-II DLL

to unity, the same as the type-I DLL. This property is widely utilized in a source-synchronous (or mesochronous) serial link to cancel out relative jitter between the input (d_1/d_2) and the output (d_{out}).

7.2.2 Jitter transfer of VCDL jitter and PD/CP noise

Because a DLL has multiple internal jitter generation sources, jitter transfer functions of the internally generated jitters should be considered, in addition to the transfer functions of the input jitter sources derived in (7.3), (7.9), and (7.10). We will consider two main jitter sources in DLLs: one is the jitter induced in the VCDL and the other is that induced by the phase detector (including the charge pump). The examples of the jitter from the VCDL supply noise and the phase detector dithering are described in Figure 7.8. For the first example, the supply voltage in a chip can never be clean. Typically, the delay of a delay cell decreases as the supply voltage increases, and therefore the supply noise introduces a delay variation to the VCDL. This delay variation shows up at the output of the VCDL even if the input

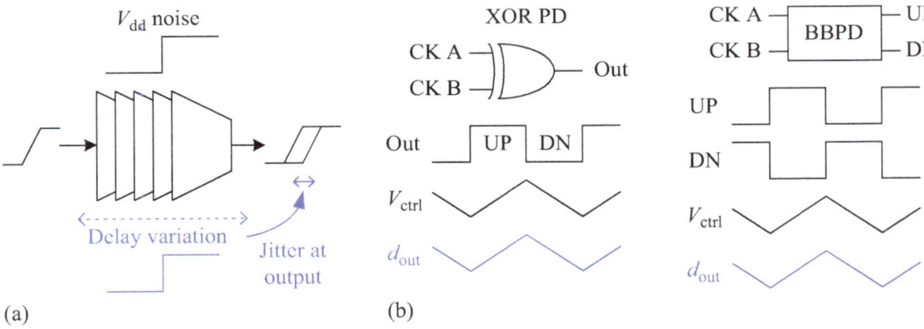

Figure 7.8 Jitter induced by (a) the VCDL and (b) the phase detector

of the VCDL does not contain jitter. The second example shows the dithering of the phase detector. Depending on the phase detector topology, such as XOR phase detector or bang-bang phase detector, the up and down pulses are repeated alternately in the steady state hence making the phase detector average output be zero across multiple cycles. However, for every single cycle, the control voltage rises and falls repeatedly which is converted to the fast fluctuation on the VCDL delay.

As observed in the example in Figure 7.8, instantaneous delay variation at the VCDL causes jitter at the output. Therefore, the VCDL jitter (d_n) can be modeled to be added at the output of the VCDL as shown in Figure 7.9. Similar to the previous derivations, d_{in} is assumed to be zero, which leads to that only the d_{out} contributes to the control voltage ($-d_{out}K_{PDCP}/sC$). As a result, the sum of the feedback term from d_{out} and d_n equals to d_{out}. By solving the equation, a high-pass transfer function is obtained as

$$H(s) = \frac{d_{out}}{d_n} = \frac{s}{s + K_{PDCP}K_{VCDL}/C} \tag{7.11}$$

The transfer function of the phase detector noise is derived in Figure 7.10. Because the current noise or the offset of the charge-pump is another important jitter source in a DLL, the noise is added at the output of the charge-pump in the phase domain model, to take into account the jitter contribution of the charge pump. The transfer function is calculated as

$$H(s) = \frac{d_{out}}{d_n} = \frac{K_{VCDL}/C}{s + K_{PDCP}K_{VCDL}/C} \tag{7.12}$$

Contrary to the VCDL jitter, (7.12) shows a low-pass characteristic. Here we can find the trade-off of the DLL design; the jitter induced by the VCDL is high-pass filtered but that induced by the PD/CP is low-pass filtered, with respect to the loop bandwidth. Therefore, the loop bandwidth should be chosen carefully to minimize the overall jitter as like in PLL designs.

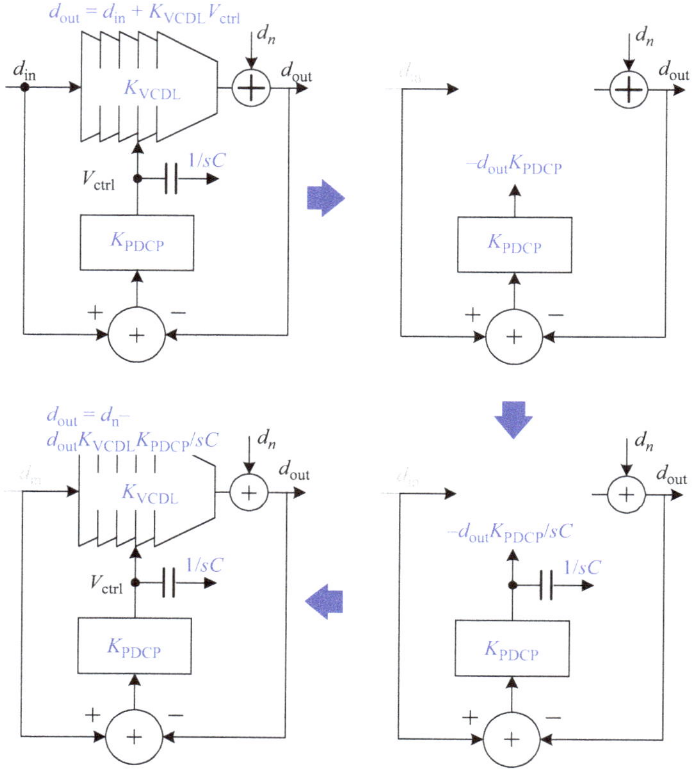

Figure 7.9 Derivation of VCDL jitter transfer function

7.3 Jitter generation and transfer of open-loop clock buffer

Through Chapter 6 and this chapter, we studied the jitter transfer functions of various jitter sources in PLL/DLLs; for example, the input reference jitter is low-pass filtered after passing through a PLL but is not filtered through a DLL. On the other hand, the jitter generated from VCO/VCDL is high-pass filtered by the feedback loop. Then what happens to the input jitter when it passes just a simple buffer stage? Also, the readers may wonder how the jitter generation by the buffer stage affects the jitter profile. In this chapter, we examine the jitter generation and jitter transfer from clock buffer stage without feedback.

Figure 7.11 shows a simple CMOS clock buffer, where i_n represents the noise current of the CMOS transistors. As we derived in the ISF of a ring oscillator, the additive noise current affects the timing only when the noise injection happens during the first half of an edge transition, whose duration equals to the delay of the

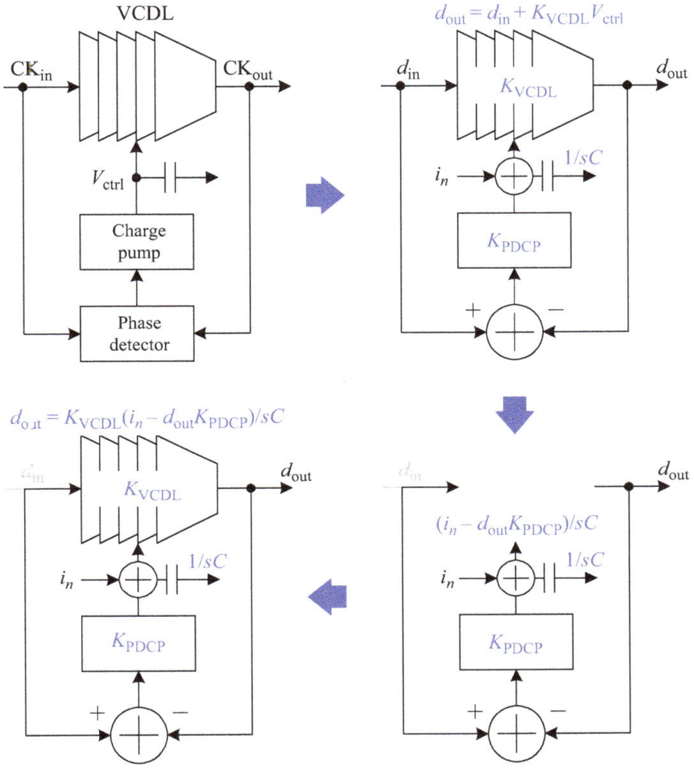

Figure 7.10 Derivation of phase detector and charge-pump noise transfer function

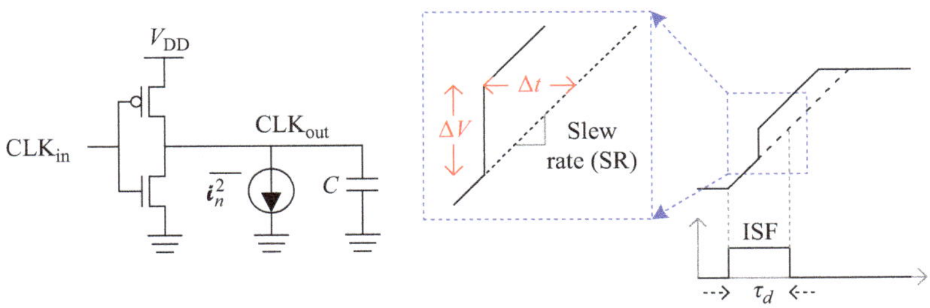

Figure 7.11 CMOS clock buffer with noise current and jitter induced by the noise

inverter (τ_d) assuming the first-order approximation and relatively fast input transition. The amount of the time shift is expressed as

$$\Delta t = \frac{\Delta V}{\text{SR}} = \frac{1}{\text{SR}} \cdot \frac{\Delta Q}{C} = \frac{1}{\text{SR}} \cdot \frac{1}{C} \cdot \int_0^{\tau_d} i_n(t)dt \tag{7.13}$$

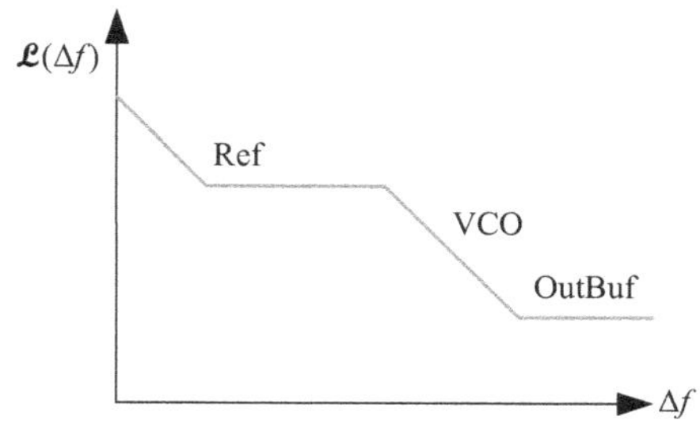

Figure 7.12 Generic phase noise profile of practical measurement

where SR is the slew rate of the transition. Typically, the τ_d is much less than the period of the clock (T_{clk}), for example, as a rule of thumb $\tau_d < T_{\mathrm{clk}}/8$. As a result, we can approximate that $i_n(t)$ is a constant during the short time of τ_d, if we assume that higher frequency components are filtered out by the limited bandwidth of the buffer. Then (7.13) is simplified to

$$\Delta t \cong \frac{1}{\mathrm{SR}} \cdot \frac{1}{C} \cdot \tau_d \cdot i_n = \frac{V_{\mathrm{DD}}}{2 \cdot \mathrm{SR}^2} \cdot \frac{1}{C} \cdot i_n \tag{7.14}$$

where we find that the jitter is inversely proportional to the square of the slew rate, so it is important to keep a fast slew rate to minimize the jitter generation by a buffer. As a rule of thumb, fanout of no more than 4 is highly recommended for sensitive signals. In addition, we can also observe that the PSD of Δt follows the same profile as the noise current, which means that it will be a white jitter if we consider the white noise from the transistors. In fact, $1/f$ noise also affects, however, buffer stages are generally wideband, so the contribution from the $1/f$ noise is negligible compared to the white noise. Because of its frequency-independency, it dominates the other jitter components at very high frequency even though its noise floor is very low. As a result, a practical phase noise measurement from a CMOS clock generation circuit becomes like Figure 7.12, because any kind of output buffer stage should be used to drive circuit load and to drive the clock output to the test equipment.

In order to identify the jitter transfer of a buffer stage, we can examine how the jitter is affected when a jittery clock signal passes through a filter (Figure 7.13). The input clock signal can be expressed as

$$\mathrm{in}(t) = \sin(2\pi f_0 t) + a \cdot \sin(2\pi (f_0 + \Delta f)t) + a \cdot \sin(2\pi (f_0 - \Delta f)t) \tag{7.15}$$

where f_0 and Δf are the clock frequency and the jitter frequency. Note that the amplitude of the fundamental signal is normalized for simplicity. From the definition, the phase noise is calculated as

$$\mathcal{L}_{\mathrm{in}}(\Delta f) = 10 \log\left(a^2\right) \tag{7.16}$$

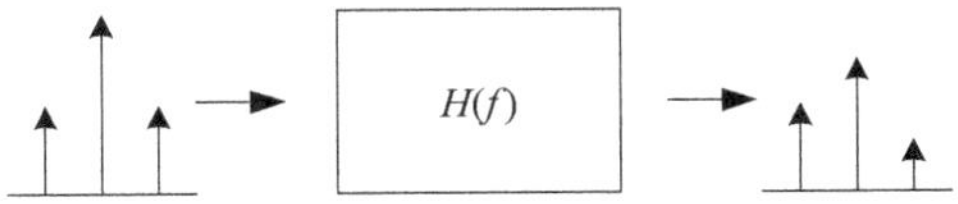

Figure 7.13 Filter with jittery input

On the other hand, assuming the phase delay is a constant across the frequency range of interest [3], the output from the filter is expressed with the magnitude response of the filter as

$$\begin{aligned}\text{out}(t) &= H(f_0)\sin(2\pi f_0 t) + aH(f_0+\Delta f)\cdot\sin(2\pi(f_0+\Delta f)t)\\&\quad + aH(f_0-\Delta f)\cdot\sin(2\pi(f_0-\Delta f)t)\end{aligned} \tag{7.17}$$

The phase noise is obtained as

$$\mathcal{L}(\Delta f) = 10\log\left(\frac{a^2}{2}\cdot\frac{|H(f_0+\Delta f)|^2+|H(f_0-\Delta f)|^2}{|H(f_0)|^2}\right) \tag{7.18}$$

In general, a single buffer stage exhibits a simple low-pass characteristic as

$$H(f) = \frac{1}{1+\frac{jf}{f_c}} \tag{7.19}$$

where f_c is the cut-off frequency of the filter. By substituting (7.19) into (7.18), we obtain:

$$\mathcal{L}_{\text{out}}(\Delta f) = 10\log\left(\frac{a^2}{2}\cdot\left(\frac{1+\left(\frac{f_0}{f_c}\right)^2}{1+\left(\frac{f_0+\Delta f}{f_c}\right)^2}+\frac{1+\left(\frac{f_0}{f_c}\right)^2}{1+\left(\frac{f_0-\Delta f}{f_c}\right)^2}\right)\right) \tag{7.20}$$

From (7.16) and (7.20), the jitter transfer function is achieved as

$$\begin{aligned}J_{\text{LPF}}(\Delta f) &= \mathcal{L}_{\text{out}}(\Delta f) - \mathcal{L}_{\text{in}}(\Delta f)\\&= 10\log\left(\frac{1}{2}\cdot\left(\frac{1+\left(\frac{f_0}{f_c}\right)^2}{1+\left(\frac{f_0+\Delta f}{f_c}\right)^2}+\frac{1+\left(\frac{f_0}{f_c}\right)^2}{1+\left(\frac{f_0-\Delta f}{f_c}\right)^2}\right)\right)\end{aligned} \tag{7.21}$$

Jitter transfer functions for various cut-off frequencies obtained from (7.21) is plotted in Figure 7.14. As long as the cut-off frequency is less than the fundamental frequency, we can find that the high-frequency jitter is amplified after the clock signal passes a low-pass filter, which is referred to as jitter amplification [4]. For example, if the cut-off frequency of the low-pass filter is 1/4 of the clock frequency, the jitter is amplified up to 9 dB depending on the frequency. For the cut-off

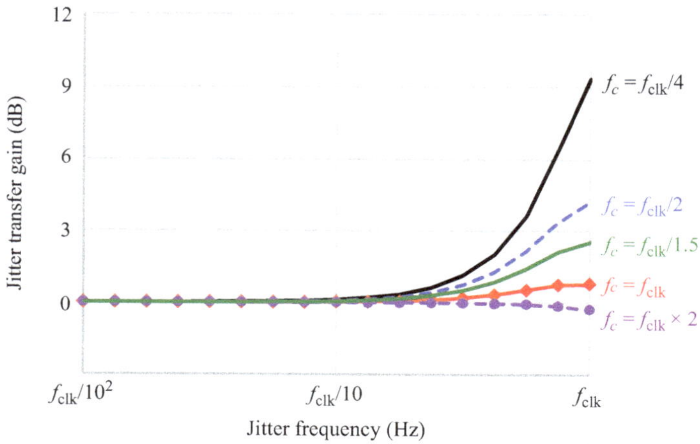

Figure 7.14 Jitter transfer functions of low-pass filter for various cut-off frequency

frequency higher than ~$0.6f_{clk}$, we can limit the jitter amplification less than 3 dB for all frequency. Here, note that only a single-stage buffer is considered in, but the jitter amplification is accumulated exponentially through a buffer chain. Therefore, it is very important to retain a sufficient bandwidth for a buffer stage, if a buffer chain is used to drive a highly precise clock signal.

We can also examine duty-cycle error amplification, which is another important metric while designing a clock buffer, using (7.21). In fact, the duty cycle distortion is denoted as a DC component of jitter, so (7.21) can be simplified by assuming $\Delta f = f_0$ and neglecting the $(f_0 + \Delta f)$ term:

$$J_{DCD} = 10\log\left(1 + \left(\frac{f_0}{f_c}\right)^2\right) \tag{7.22}$$

Equation (7.22) is plotted in Figure 7.15 for various cut-off frequencies of the low-pass filter, with respect to the input clock frequency. Intuitively, the fundamental signal experiences the attenuation depending on the cut-off frequency and the fundamental frequency whereas there is no attenuation to the duty-cycle distortion; hence the duty-cycle distortion grows inversely proportional to the filter attenuation at the fundamental frequency. As a rule of thumb, the bandwidth of a buffer should be higher than the clock frequency to suppress the amplification less than 3 dB.

On the other hand, when the clock signal passes a high-pass filter, a different aspect is observed. For a high-pass filter, we can substitute:

$$H(f) = \frac{\frac{jf}{f_c}}{1 + \frac{jf}{f_c}} \tag{7.23}$$

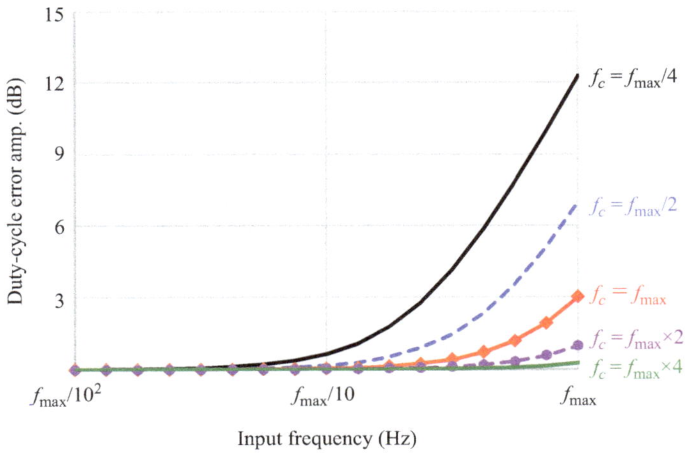

Figure 7.15 Duty-cycle error amplification for various cut-off frequencies with respect to the input frequency

into (7.18) and obtain:

$$\mathcal{L}_{out}(\Delta f) = 10\log\left(\frac{a^2}{2}\cdot\left(\left(\frac{f_0+\Delta f}{f_0}\right)^2\cdot\frac{1+\left(\frac{f_0}{f_c}\right)^2}{1+\left(\frac{f_0+\Delta f}{f_c}\right)^2}+\left(\frac{f_0-\Delta f}{f_0}\right)^2\cdot\frac{1+\left(\frac{f_0}{f_c}\right)^2}{1+\left(\frac{f_0-\Delta f}{f_c}\right)^2}\right)\right) \tag{7.24}$$

From (7.16) and (7.24), we obtain the jitter transfer function of a high-pass filter to

$$J_{HPF}(\Delta f) = 10\log\left(\frac{1}{2}\cdot\left(\left(\frac{f_0+\Delta f}{f_0}\right)^2\cdot\frac{1+\left(\frac{f_0}{f_c}\right)^2}{1+\left(\frac{f_0+\Delta f}{f_c}\right)^2}+\left(\frac{f_0-\Delta f}{f_0}\right)^2\cdot\frac{1+\left(\frac{f_0}{f_c}\right)^2}{1+\left(\frac{f_0-\Delta f}{f_c}\right)^2}\right)\right) \tag{7.25}$$

Equation (7.25) is illustrated in Figure 7.16. If the cut-off frequency is fairly higher than the clock frequency, the high-frequency jitter is amplified similar to the low-pass filter. On the other hand, for a low cut-off frequency, where we can find more practical examples in high-speed clocking circuits, the jitter transfer function becomes low-pass in contrast to the low-pass filter case. That is, interestingly, the jitter is low-pass filtered when the clock signal passes a high-pass filter.

On the other hand, the duty-cycle error is corrected by a high-pass filter. Intuitively, since the duty-cycle error is DC component, the high-pass filter suppresses the duty-cycle error propagating to the output. As a result, a simple high-pass filter implementation of AC-coupled buffer shown in Figure 7.17 is frequently used in many applications [5–9]. The AC-coupling capacitor blocks the

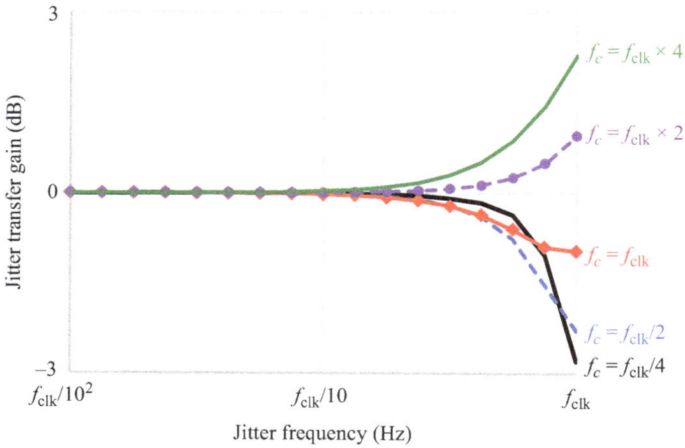

Figure 7.16 Jitter transfer functions of high-pass filter for various cut-off frequency

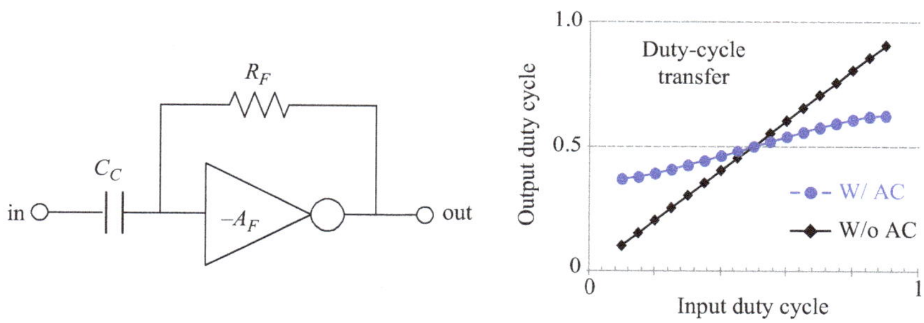

Figure 7.17 AC-coupled buffer and duty-cycle error transfer of the buffer

low-frequency component of the input, and the feedback resistor sets common-mode voltage to the crossover voltage. With Miller approximation, the high-pass cut-off frequency is simplified to

$$f_c = \frac{1 + A_F}{2\pi R_F C_C} \tag{7.26}$$

A robust implementation of duty-cycle correction circuit is shown in Figure 7.18. It adopts a negative feedback which detects the duty-cycle error at the output of buffer chain and corrects it by adjusting the input common mode. Knowing that the duty-cycle is the DC component of clock, a low-pass filter is used to extract the duty-cycle information from the output. An error amplifier compares the extracted voltage to the reference voltage which is supposed to be an ideal crossover voltage. The error output is translated to current which controls the input common mode by charging or discharging the input node. A linearized diagram of

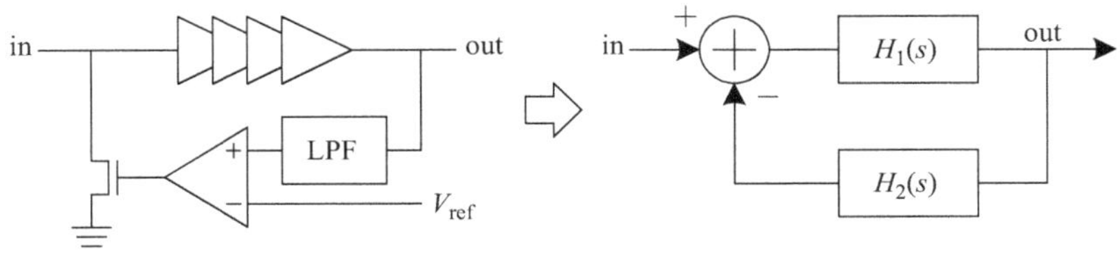

Figure 7.18 Simplified circuit diagram and linear model of feedback-based duty-cycle correction

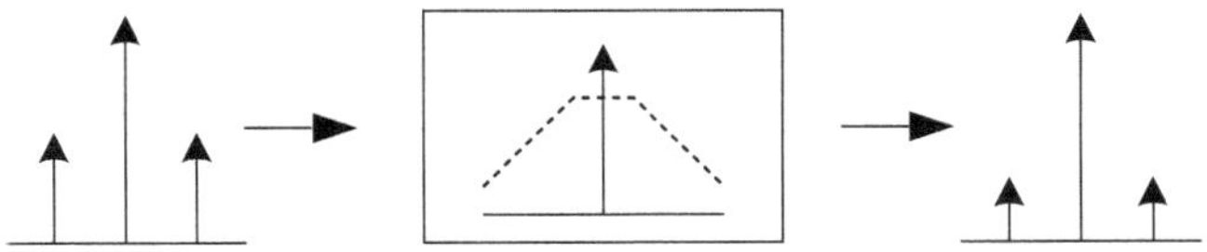

Figure 7.19 Jitter transfer of a band-pass filter

such duty-cycle correction is shown on the right side of Figure 7.18, where $H_1(s)$ and $H_2(s)$ represent the transfer functions of the buffer chain and the entire feedback path. Simplifying $H_1(s)$ and $H_2(s)$ to first-order low-pass filters gives closed-loop transfer function as

$$H(s) = \frac{H_1(s)}{1 + H_1(s)H_2(s)} = \frac{\frac{1}{1+s/\omega_1}}{1 + \frac{k}{(1+s/\omega_1)(1+s/\omega_2)}} = \frac{1+\frac{s}{\omega_2}}{\left(1+\frac{s}{\omega_1}\right)\left(1+\frac{s}{\omega_2}\right)+k} \tag{7.27}$$

where k is the feedback gain. Of course ω_2 should be much lower than ω_1. We can find a zero at ω_2, which implies that (7.27) exhibits high-pass transfer before ω_1.

A band-pass filter case can be understood more intuitively. With a well-tuned band-pass filter, the unwanted frequency component (jitter) is suppressed by the filter while the fundamental signal is not attenuated, as shown in Figure 7.18. In other words, the jitter is low-pass filtered by a band-pass filter. Because the band-pass filter suppresses the jitter component at both sidebands, it reduces jitter further than a high-pass filter. In fact, due to the limited output bandwidth, the AC-coupled buffer in Figure 7.19 is a band-pass filter, so the buffer reduces the jitter considerably with careful design.

7.4 Design consideration on number of stages and tuning range of DLL

Similar to ring VCOs, the number of stages of delay line is one of the most critical design parameters in DLL designs because it determines tuning range. Simply, the number of stages can be chosen among the integer numbers which meet the following criterion:

$$N \cdot \tau_{\min} < T < N \cdot \tau_{\max} \tag{7.28}$$

where $\tau_{\min}$ and $\tau_{\max}$ are the minimum and maximum delays of a single stage within the tuning range, and T is the desired delay at proper lock. At the same time, the tuning range of DLL is written as

$$\mathrm{DR} = N \cdot (\tau_{\max} - \tau_{\min}) \tag{7.29}$$

Considering the PVT variations, (7.28) can be rewritten as

$$N \cdot \tau_{\min,\text{slow}} < T < N \cdot \tau_{\max,\text{fast}} \tag{7.30}$$

where $\tau_{\min}$ and $\tau_{\max}$ are the minimum delay at the slowest PVT corner and the maximum delay at the fastest corner, respectively. Figure 7.20 implies that (7.30) is much hard to satisfy compared to (7.28). Hence, dynamic range is generally required to be wide to cover the desired delay across the PVT corners. Further wider range is needed if a DLL covers a wide frequency range. Therefore, it is worthwhile to see what sets the $\tau_{\min}$ and the $\tau_{\max}$. The $\tau_{\min}$ is a matter of how fast the transient response of the delay cell is, as a result, it is typically limited by the process technology and power budget. On the other hand, the $\tau_{\max}$ considers different aspects. Unlike ring VCOs where the minimum frequency (maximum delay) of the tuning range does not matter much, a VCDL is sensitive to the maximum delay. It is mainly because a ring oscillator creates clock by itself; however, a delay line should be provided an input clock from another source. At the maximum delay, the ring oscillator produces a slow clock coinciding with the delay, so there is no bandwidth issue. However, the bandwidth of the delay line should be carefully considered because the delay line always operates at the input frequency, especially at the maximum delay where the circuit bandwidth is minimized. For an extreme example, if the circuit bandwidth of delay line is far less than the clock frequency at

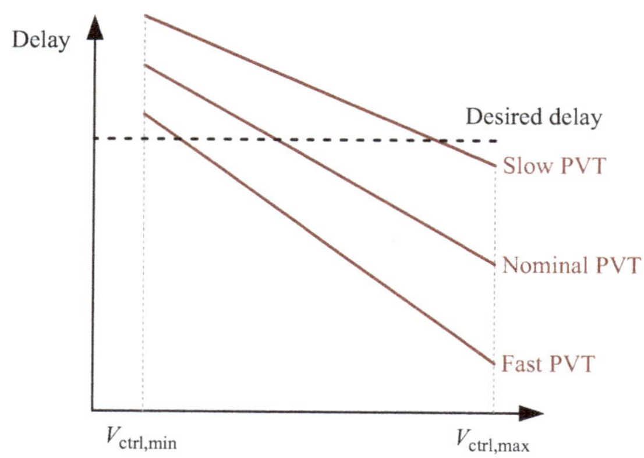

Figure 7.20 Dynamic range of DLL (VCDL) across PVT variations

the maximum delay but the DLL initially wakes up at the maximum delay, the input clock may disappear while passing through the delay line. Then phase detector does not produce any output since there is no feedback clock, leading to a global convergence failure of DLL. Generally speaking, the bandwidth should be kept above a certain level not to result in any malfunction due to the jitter or duty-cycle amplification that we studied in Section 7.3. This criterion can be written as

$$f_{\mathrm{BW,min}} > k \cdot f_{\mathrm{clk}} \tag{7.31}$$

where k is a constant set by the jitter and duty-cycle amplification requirement. Assuming a simple RC delay model for the buffer, we can substitute the following relations:

$$\tau_{\mathrm{max}} = \ln 2 \cdot R_{\mathrm{max}} C \tag{7.32}$$

$$f_{\mathrm{BW,min}} = \frac{1}{2\pi R_{\mathrm{max}} C} \tag{7.33}$$

into (7.31), and then obtain:

$$\tau_{\mathrm{max}} < \frac{\ln 2}{2\pi k f_{\mathrm{clk}}} \tag{7.34}$$

With $k \approx 1$, (7.34) ends up by the rule-of-thumb of $f_{\mathrm{clk}} < 1/8\tau$ [10]. To summarize, given the process parameters, the minimum and maximum delays of a delay unit are, respectively, designed from the power budget and the bandwidth requirement, and the number of stages is chosen accordingly to include the desired delay in the tuning range across the PVT variations.

References

[1] Yang CK. Delay-locked loops-an overview. *Phase-Locking in High-Performance Systems*. New York, NY: Wiley-IEEE Press; 2003. pp. 13–22.

[2] Lee MJ, Dally WJ, Greer T, *et al.* Jitter transfer characteristics of delay-locked loops-theories and design techniques. *IEEE Journal of Solid-State Circuits*. 2003;38(4):614–621.

[3] Bae W, Nikolić B, Jeong DK. Use of phase delay analysis for evaluating wideband circuits: An alternative to group delay analysis. *IEEE Transactions on Very Large Scale Integration (VLSI) Systems*. 2017;25(12): 3543–3547.

[4] Casper B, O'Mahony F. Clocking analysis, implementation and measurement techniques for high-speed data links: A tutorial. *IEEE Transactions on Circuits and Systems I: Regular Papers*. 2009;56(1):17–39.

[5] Bae W. CMOS inverter as analog circuit: An overview. *Journal of Low Power Electronics and Applications*. 2019;9(3):26.

[6] Chen S, Zhou L, Zhuang I, *et al.* A 4-to-16GHz inverter-based injection-locked quadrature clock generator with phase interpolators for multi-standard

I/Os in 7nm FinFET. In *2018 IEEE International Solid-State Circuits Conference-(ISSCC)*. San Francisco, CA: IEEE; 2018 (pp. 390–392).

[7] Wang H, Lee J. A 21-Gb/s 87-mW transceiver with FFE/DFE/analog equalizer in 65-nm CMOS technology. *IEEE Journal of Solid-State Circuits.* 2010;45(4):909–920.

[8] Song YH, Bai R, Hu K, Yang HW, Chiang PY, Palermo S. A 0.47–0.66 pJ/bit, 4.8–8 Gb/s I/O transceiver in 65 nm CMOS. *IEEE Journal of Solid-State Circuits.* 2013;48(5):1276–1289.

[9] Savoj J, Hsieh KC, An FT, *et al.* A low-power 0.5–6.6 Gb/s wireline transceiver embedded in low-cost 28 nm FPGAs. *IEEE Journal of Solid-State Circuits.* 2013;48(11):2582–2594.

[10] Lee MJ, Dally WJ, Chiang P. Low-power area-efficient high-speed I/O circuit techniques. *IEEE Journal of Solid-State Circuits.* 2000;35(11): 1591–1599.

Chapter 8

Phase noise suppression techniques 1: subsampling PLL

8.1 Introduction

In Chapter 6, we studied the phase noise contributions in a PLL; the VCO noise dominates at frequency region higher than the PLL loop bandwidth, but the input and the phase detector and charge-pump (PD/CP) noise dominates the in-band noise, so there is a trade-off. We also studied how to find the optimum cut-off frequency. In Chapter 4, on the other hand, we studied the phase noise of CMOS oscillators. If all parameters (i.e., Q of inductor and rise/fall time of ring oscillator) are designed well, the only way to improve the phase noise is increasing power consumption (recall (4.8) and (4.58). This criterion is always true for any analog circuit. In other words, once we find the optimum bandwidth, there is nothing to do except for burning more power to suppress the phase noise. The motivation of subsampling PLL (SS-PLL) is to tackle this old criterion in PLL. In specific, reducing the in-band noise contribution from the PD/CP without consuming additional power is the motivation of SS-PLL.

Recalling (6.7), the closed-loop transfer function of a PLL is written as

$$H(s) = \frac{T(s)}{1 + T(s)/N} \tag{8.1}$$

Dividing (8.1) by the transfer function of PD/CP leads to noise transfer function from the PD/CP noise as

$$\frac{\Delta\phi_{\text{out}}}{\Delta i_{\text{PD,CP}}} = \frac{2\pi}{I_{\text{CP}}} \cdot \frac{T(s)}{1 + T(s)/N} \tag{8.2}$$

$$\mathcal{L}(\Delta f) = 10\log\left(\frac{2kTF}{P_{sig}}\left(\frac{f_0}{2Q\Delta f}\right)^2 + \frac{2kT\gamma g_m R}{P_{sig}}\left(\frac{f_0}{2Q\Delta f}\right)^2\left(\frac{f_{1/f}}{\Delta f}\right)\right) \tag{4.8}$$

$$\mathcal{L}(\Delta\omega) = 10\log\left(\frac{8kT\gamma}{I}\cdot\frac{g_m}{I}\cdot\frac{\omega_0^2}{\Delta\omega^2}\right) = 10\log\left(\frac{8kT\gamma}{I}\cdot\frac{2}{V_{OV}}\cdot\frac{\omega_0^2}{\Delta\omega^2}\right) \tag{4.58}$$

While considering the in-band noise, we can assume that $T(s) >> 1$. Then (8.2) is simplified to

$$\frac{\Delta\phi_{\mathrm{out}}}{\Delta i_{\mathrm{PD,CP}}} \cong \frac{2\pi}{I_{\mathrm{CP}}} \cdot N \tag{8.3}$$

Substituting (8.3) into the phase noise relation to the phase modulation (2.23), we can obtain the in-band phase noise contributed from the PD/CP as

$$\begin{aligned}\mathcal{L}_{\mathrm{PD,CP}}(\Delta f) &= 10\log\left(\frac{S_{\Delta\phi_{\mathrm{out}}}(\Delta f)}{2}\right)\\ &\cong 10\log\left(\left(\frac{2\pi}{I_{\mathrm{CP}}}\cdot N\right)^2 \cdot \frac{S_{\Delta i_{\mathrm{PD,CP}}}(\Delta \mathrm{f})}{2}\cdot\frac{t_{\mathrm{PFD}}}{T_{\mathrm{ref}}}\right)\end{aligned} \tag{8.4}$$

where t_{PFD} is the PFD reset delay. Note that the multiplication factor of $\frac{t_{\mathrm{PFD}}}{T_{\mathrm{ref}}}$ is introduced to reflect the fact that the charge-pump flows current only for the t_{PFD} during normal steady-state operation. From (8.4), we can find that the in-band phase noise is amplified by a factor of N^2, which is the main challenge that a subsampling PLL tackles:

8.2 Subsampling PLL

The main idea of subsampling PLL is to remove the divider in the PLL feedback path for eliminating the noise amplification by the division factor N [1]. If the divider is removed, the phase detector should compare the reference clock and the VCO clock, whose frequencies are much different even when the PLL is locked. None of the phase detectors we studied in Chapter 5 offers such operation. Reference [2] proposed a subsampling phase detector (SSPD) which is implemented with a sample-and-hold circuit followed by a V–I converter. Figure 8.1 explains the operation of the SSPD. The sample-and-hold circuit tracks the VCO clock while the reference clock is at low but holds the previous state when the reference clock is at high. As long as the sampling switch has a sufficient bandwidth, the sampling capacitor stores the voltage level of the VCO clock at the moment of rising edge of the reference clock. The g_{m} element followed by the sample-and-hold circuit converts the sampled voltage to the current, $g_{\mathrm{m}}V_{\mathrm{samp}}$, which can be processed with a conventional current-driven loop filter. If the VCO leads the reference clock, the sampling happens when the VCO clock is higher than its common-mode voltage ($V_{\mathrm{CM,VCO}}$). Assuming that the VCO has a positive K_{VCO} (i.e., frequency increases as the control voltage increases), the control voltage needs to be lowered. Therefore, the $g_{\mathrm{m}}V_{\mathrm{samp}}$ is used as a pull-down current. The pull-up

---

$$\mathcal{L}(\Delta \mathrm{f}) = 10\log\left(\frac{1}{2}S_{\phi}(\Delta \mathrm{f})\right) \tag{2.23}$$

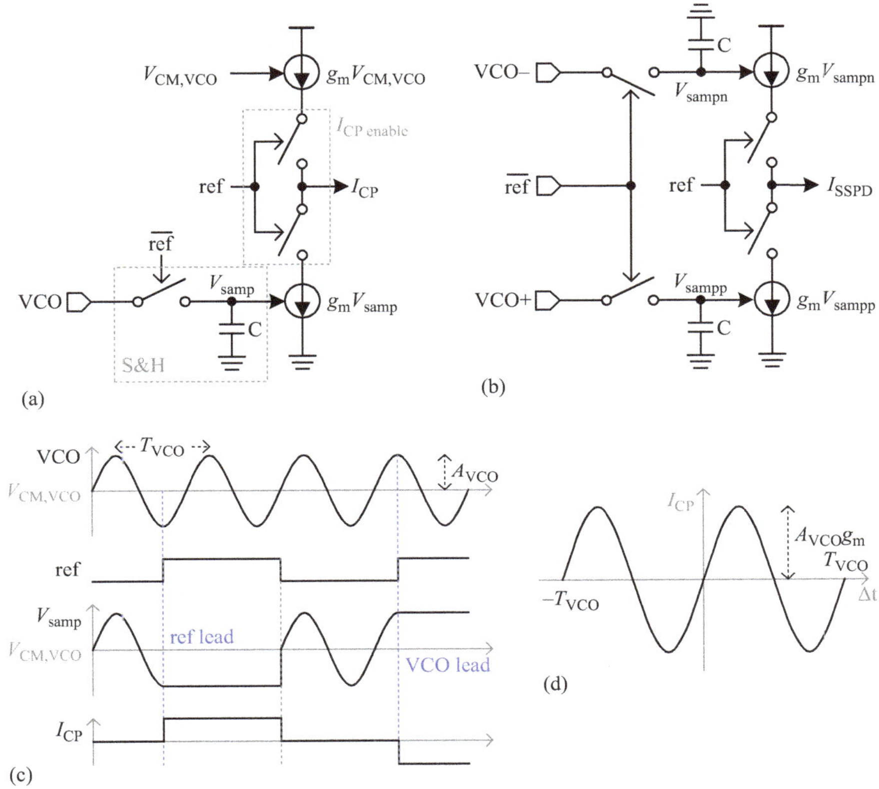

Figure 8.1. Subsampling phase detector: (a) simplified circuit diagram, (b) differential configuration, (c) waveforms, and (d) gain curve

current is obtained from the $V_{CM,VCO}$ to nullify the total current under locked. I_{CP} enabling switches are placed to isolate the current flowing during the sampling cycle. Therefore, those I_{CP} switches turn on complementarily to the sampling switch, so the combination of the enabling switch and the sampling switch can be regarded as an analog master-slave flip-flop. As a result, the charge-pump current (I_{SSPD}) flows only when the SSPD is in the hold cycles. An example of voltage and current waveforms are shown in Figure 8.1(c). A differential configuration shown in Figure 8.1(b) prevents any non-ideal effects introduced from the $V_{CM,VCO}$ but also provides a better supply noise rejection.

The primary difference over the conventional phase detector and charge-pump is that the SSPD relies on the amplitude and waveform of the VCO clock. Recall that the amount of pumped charge is time-controlled by turning on the pump only when there is a time difference, which is extracted only from edge information at the phase detector, while the pump currents are fixed in the conventional PLL. In the SSPD, the current is proportional to the sampled voltage whereas the turn-on time is fixed. As a result, the SSPD gain curve has the same shape as the waveform of the VCO clock, as shown in Figure 8.1(d) where the waveform is assumed to be a sine wave. The SSPD gain at the locking point is calculated as

$$\left.\frac{\Delta i_{\mathrm{SSPD}}}{\Delta\phi_{\mathrm{err}}}\right|_{\Delta\phi_{\mathrm{err}}=0} = \frac{1}{2\pi f_{\mathrm{VCO}}}\cdot\left.\frac{\Delta i_{\mathrm{SSPD}}}{\Delta t_{\mathrm{err}}}\right|_{\Delta t_{\mathrm{err}}=0} = g_m A_{\mathrm{VCO}} \quad (8.5)$$

Note that the gain is not a function of N since the SSPD compares the phase at f_{VCO}, rather than f_{ref}. The in-band phase noise from the SSPD becomes:

$$\mathcal{L}_{\mathrm{SSPD}}(\Delta \mathrm{f}) \cong 10\log\left(\left(\frac{1}{g_m A_{\mathrm{VCO}}}\right)^2\cdot\frac{\mathrm{S}_{\Delta i_{\mathrm{SSPD}}}(\Delta f)}{2}\right) \quad (8.6)$$

From (8.4) and (8.6), we can compare the in-band noise from the conventional PFD/CP and the SSPD as

$$\frac{\mathcal{L}_{\mathrm{SSPD}}(\Delta f)}{\mathcal{L}_{\mathrm{PD,CP}}(\Delta f)} \cong 10\log\left(\left(\frac{1}{g_{m,\mathrm{SSPD}}A_{\mathrm{VCO}}}\right)^2\cdot\left(\frac{I_{CP}}{2\pi N}\right)^2\cdot\frac{T_{ref}}{t_{PFD}}\cdot\frac{\mathrm{S}_{\Delta i_{\mathrm{SSPD}}}(\Delta \mathrm{f})}{\mathrm{S}_{\Delta \mathrm{i}_{\mathrm{PD,CP}}}(\Delta \mathrm{f})}\right) \quad (8.7)$$

Assuming the transconductance is the same for the pull-up and the pull-down, (8.7) is simplified as

$$\frac{\mathcal{L}_{\mathrm{SSPD}}(\Delta f)}{\mathcal{L}_{\mathrm{PD,CP}}(\Delta f)} \cong 10\log\left(\left(\frac{1}{g_{m,\mathrm{SSPD}}A_{\mathrm{VCO}}}\right)^2\cdot\left(\frac{I_{\mathrm{CP}}}{2\pi N}\right)^2\cdot\frac{T_{\mathrm{ref}}}{t_{\mathrm{PFD}}}\cdot\frac{g_{m,\mathrm{SSPD}}}{g_{m,\mathrm{CP}}}\right) \quad (8.8)$$

which can be re-written as

$$\frac{\mathcal{L}_{\mathrm{SSPD}}(\Delta f)}{\mathcal{L}_{\mathrm{PD,CP}}(\Delta f)} \cong 10\log\left(\left(\frac{1}{g_{m,\mathrm{SSPD}}A_{\mathrm{VCO}}}\right)^2\cdot\left(\frac{I_{CP}}{2\pi N}\right)^2\cdot\frac{T_{\mathrm{ref}}}{t_{\mathrm{PFD}}}\cdot\frac{I_{\mathrm{SSPD}}}{I_{\mathrm{CP}}}\right) \quad (8.9)$$

if we assume the same gate-overdrive voltage for the I_{CP} and the I_{SSPD} for simplicity. For a clearer comparison, here we make another assumption that the loop gain is the same for the conventional PLL and the SS-PLL. That means the SSPD gain should be equal to the PFD/CP gain to keep the same loop filter and VCO. This assumption can be expressed as

$$\frac{I_{\mathrm{CP}}}{2\pi} = g_{m,\mathrm{SSPD}}A_{\mathrm{VCO}} = \frac{2I_{\mathrm{SSPD}}}{V_{\mathrm{OV}}}\cdot A_{\mathrm{VCO}} \quad (8.10)$$

where V_{OV} represents the gate-overdrive voltage. Substituting (8.10) to (8.9) leads to

$$\frac{\mathcal{L}_{SSPD}(\Delta f)}{\mathcal{L}_{PD,CP}(\Delta f)} \cong 10\log\left(\left(\frac{1}{N}\right)^2 \cdot \frac{T_{ref}}{t_{PFD}} \cdot \frac{1}{4\pi} \cdot \frac{V_{OV}}{A_{VCO}}\right)$$

$$= 10\log\left(\frac{1}{N} \cdot \frac{T_{VCO}}{t_{PFD}} \cdot \frac{1}{4\pi} \cdot \frac{V_{OV}}{A_{VCO}}\right) \tag{8.11}$$

Note that $\frac{V_{OV}}{A_{VCO}}$ is not very flexible, and t_{PFD} is usually process-dependent because it is generally set to the minimum that avoids the dead zone. Therefore, the CP noise suppression factor is mainly dependent on N and T_{VCO}. Equation (8.11) is plotted in Figure 8.2(a) with respect to T_{VCO}/t_{PFD}. The A_{VCO} is assumed to be $2{\cdot}V_{OV}$. We can observe that the suppression is better than −20 dB within the practical range of N and T_{VCO}/t_{PFD}. The suppression becomes better with a higher N or a lower T_{VCO}, which means that we can take more benefit of SS-PLL at a higher frequency. In fact, the reference frequency from a crystal oscillator is limited to ~100 MHz. Therefore, for high-frequency applications, N is inversely proportional to T_{VCO}, so (8.11) becomes a quadratic function of T_{VCO}. This case is plotted in Figure 8.2(b).

On the other hand, the transfer function from the reference noise is still affected by N although the divider is removed from the feedback path. Because the SSPD compares phase at f_{VCO}, the input phase needs to be converted to the f_{VCO} domain from the f_{ref}, which is a reverse function of the divider. As a result, a multiplier should be placed in the reference path of the linear model of SS-PLL before the phase detector, as shown in Figure 8.3. We can easily calculate the input-to-output jitter transfer function to

$$\frac{\Delta\phi_{out}}{\Delta\phi_{ref}} = \frac{N\Delta\phi_{out}}{\Delta\phi_N} = \frac{NT(s)}{1+T(s)} \tag{8.12}$$

The low-frequency gain of (8.12) is N, which means that the in-band phase noise contributed from the reference noise is still amplified by N. The transfer function from VCO is

$$\frac{\Delta\phi_{out}}{\Delta\phi_{VCO}} = \frac{1}{1+T(s)} \tag{8.13}$$

Note that T(s) in the denominator is not divided by N in (8.12) and (8.13), unlike the conventional charge-pump PLLs. Therefore, the 3-dB bandwidth is approximated to

$$\omega_c = \frac{I_{CP}RK_{VCO}}{2} \tag{8.14}$$

On the other hand, from the gain curve of Figure 8.1(d), we can find that the SSPD does not have frequency acquisition capability. Moreover, it has possible

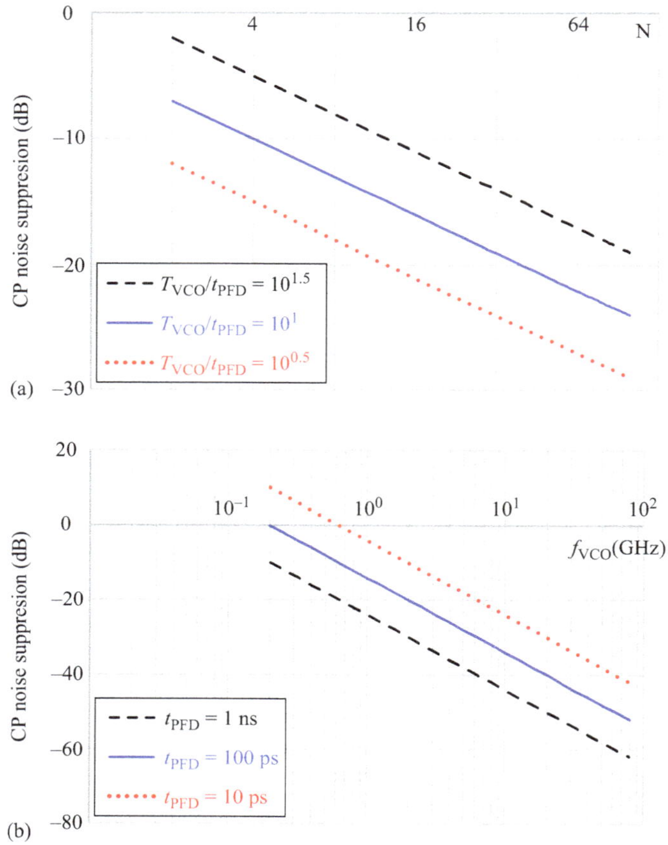

Figure 8.2. CP noise suppression of the SS-PLL: (a) a function of N while T_{VCO}/t_{PFD} is fixed and (b) a function of T_{VCO} while T_{ref}/t_{PFD} is fixed (T_{ref} = 10 ns)

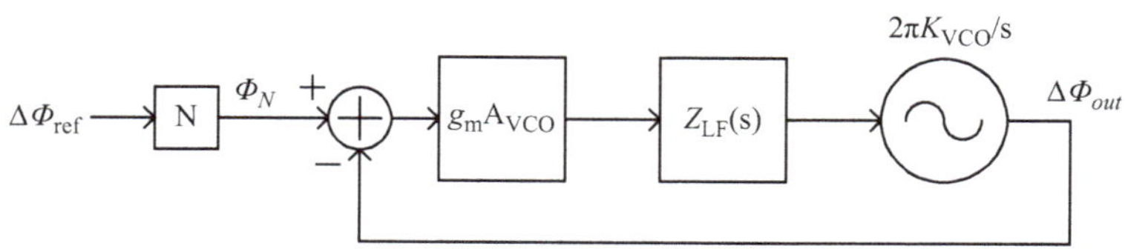

Figure 8.3. Phase domain block diagram of the SS-PLL

lock points at every T_{VCO} so that the SSPD cannot distinguish the harmonic frequencies [1]. For example, the SS-PLL can lock at $(N+1)f_{ref}$ while it is supposed to lock at Nf_{ref}, as shown in Figure 8.4(a). Therefore, an SS-PLL should work together with a frequency tracking loop, as shown in Figure 8.4(b). The phase tracking

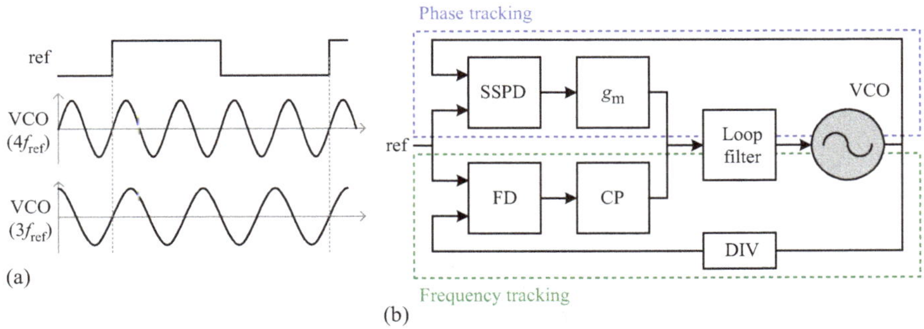

Figure 8.4. (a) Harmonic locking of SSPD and (b) SS-PLL with frequency tracking loop

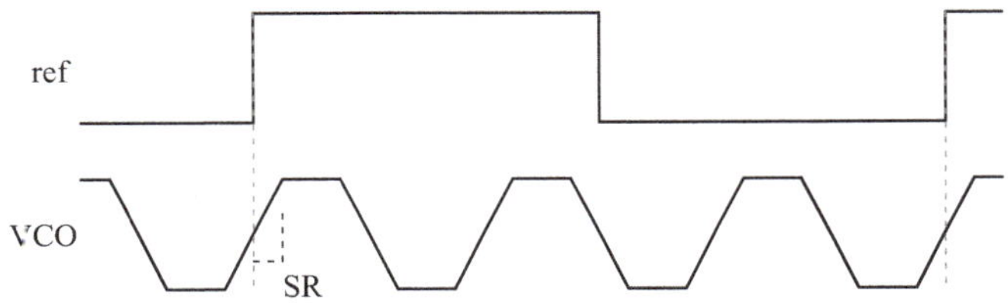

Figure 8.5. Subsampling with VCO waveform as a function of slew rate

during the steady-state should be dominated by the SS-PLL loop to utilize its superior noise suppression, whereas the frequency loop dominates while pursuing frequency acquisition. In order for that the loop gain of the frequency loop is much larger than that of the phase tracking loop. The FD should produce no output at the steady-state so that the FLL can be disabled to save power [2].

The SSPD can also be used for ring-PLLs [3–5]. As a ring-VCO does not produce a sine wave, the SSPD gain becomes a function of slew rate (V/s, Figure 8.5). The slew rate is determined by the VCO clock buffer or the number of stages of ring-VCO if the SSPD directly samples the VCO. The i_{SSPD} change by a time error (Δt_{err}) can be expressed as

$$\Delta i_{\mathrm{SSPD}} = g_m \cdot SR \cdot \Delta t_{\mathrm{err}} \tag{8.15}$$

By dividing (8.15) with $2\pi f_{\mathrm{VCO}}$, we can obtain the SSPD gain as

$$\frac{\Delta i_{\mathrm{SSPD}}}{\Delta \phi_{\mathrm{err}}} = g_m \cdot \frac{SR}{2\pi f_{\mathrm{VCO}}} \tag{8.16}$$

Because the gain is proportional to the slew rate, a tunable-slew-rate VCO buffer can be used to control the SSPD gain [6]. On the other hand, for the direct sampling on the VCO node [5], we can further simplify (8.16) using the first-order approximated waveform, which was described in Figure 4.8. Then the f_{VCO} and the slew rate are expressed with the single-stage delay (τ_d) and the number of delay stages (M) as

$$f_{\text{VCO}} = \frac{1}{2M\tau_d} \tag{8.17}$$

$$\text{SR} = \frac{V_{\text{swing}}}{2\tau_d} \tag{8.18}$$

Substituting (8.17) and (8.18), (8.16) is simplified to

$$\frac{\Delta i_{\text{SSPD}}}{\Delta \phi_{\text{err}}} = g_m V_{\text{swing}} \cdot \frac{M}{2\pi} \tag{8.19}$$

which implies that the gain is dominated by M, so a lower M is preferred to reduce the gain. In addition, the reduction of M also allows a wider linear detection range.

On the other hand, the SS-PLL is tolerant to the charge-pump mismatch. As we studied in Chapter 6, the pump-current mismatch in the conventional charge-pump PLL makes the UP/DN pulses be misaligned because the amount of charge is controlled by the pulse-width. Such misaligned UP/DN pulses introduce ripples on the control voltage causing the reference spur. In the SS-PLL, of course, there can also be a mismatch between g_m devices for UP and DN. At a steady-state, the net amount of charge injected/discharged to/from the loop filter should be zero. Since the time duration of flowing current is the same for the pull-up and the pull-down g_m, this condition can be written as

$$g_m V_{\text{CM,VCO}} = (g_m + \Delta g_m)(V_{\text{CM,VCO}} + \Delta V) \tag{8.20}$$

Equation (8.20) can be approximated to

$$|\Delta V| = \frac{\Delta g_m}{g_m} V_{\text{CM,VCO}} \tag{8.21}$$

which means that the SS-PLL introduces a voltage offset to match the UP/DN current. Note that a static phase offset is introduced for the voltage offset

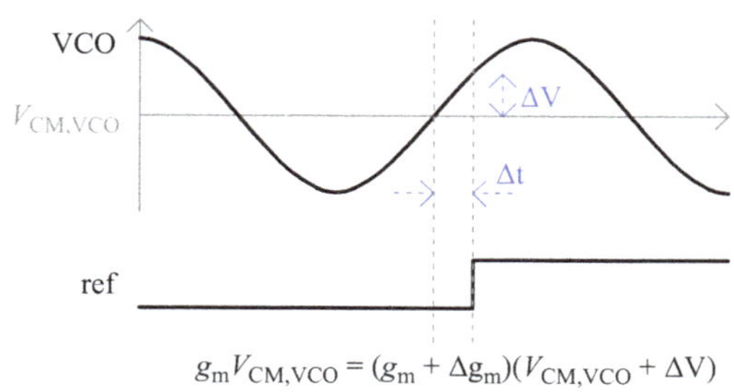

Figure 8.6. SS-PLL steady-state with g_m mismatch

(Figure 8.6); however, there is no ripple on the control voltage because the current is matched. However, the SSPD gain is reduced due to the reduced slew rate with the phase offset.

There are several possible variants for the SSPD circuit implementations, and three of which are shown in Figure 8.7. The double switch sampler SSPD moved the I_{CP} enabling switch in front of the g_m device [4, 6]. The successive complimentary samplers work like an analog master-slave flip-flop; the first one is transparent while the second one holds and vice versa. A unity-gain buffer is placed in between the samplers to prevent distortion caused by charge sharing. The reference clock pulse which turns on the I_{CP} switches, on the other hand, can be narrowed by the use of a pulse generator as shown in Figure 8.7(b). Then, they are not only slave switches but also control the SSPD gain. The output current is scaled

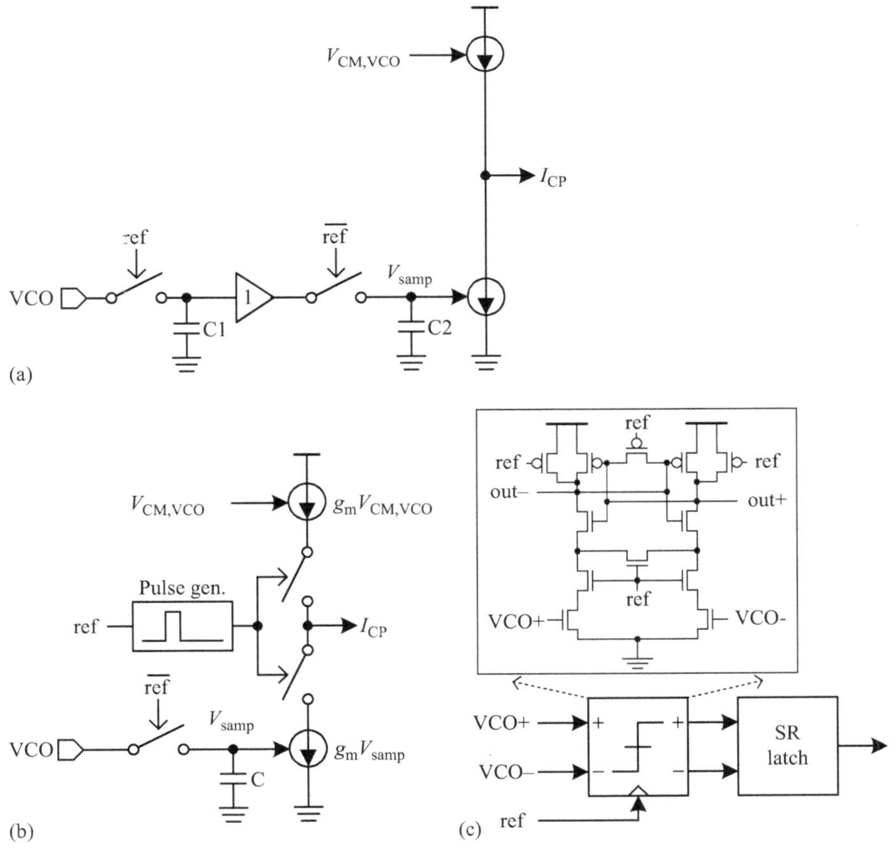

Figure 8.7. Various SSPD implementations: (a) double switch sampler [4], (b) gain control with pulse generator [2], and (c) bang-bang SSPD [5]

by the fraction of the pulse duration over the half of the reference period, and therefore the SSPD gain can be re-expressed to

$$\frac{\Delta i_{\mathrm{SSPD}}}{\Delta \phi_{\mathrm{err}}} = g_m \cdot \frac{\mathrm{SR}}{2\pi f_{\mathrm{VCO}}} \cdot \frac{2T_{\mathrm{pulse}}}{T_{\mathrm{ref}}} \tag{8.22}$$

By controlling the pulse width, the loop gain can be tuned in a wide range without affecting the SSPD operating point [1].

The subsampling technique is simply be adopted in a bang-bang system. Because it is based on the voltage sampling, a simple binary sampler, for example, a Strong Arm latch followed by an SR latch, can serve as a bang-bang SSPD [5,7]. Figure 8.7(c) shows the sense amplifier proposed in [5]. In the conventional Strong Arm latch, the input capacitance is affected by the switching of the tail NMOS, which is toggled by the clock. In the case of direct VCO sampling, it leads to fluctuation of the oscillation frequency, which is similar to the reference spur. By replacing the tail device with the three reference-clocked NMOS devices, the source of the input devices is fixed to GND so that the capacitance variation is reduced.

8.3 Fractional-N SS-PLL

Because of its detection mechanism and divider-less structure, the SS-PLL can only operate in integer-N mode inherently [8]. Although the frequency is fixed by FLL, when the VCO frequency is not an integer multiple of the reference frequency, the sampling becomes asynchronous which means the sampling phase drifts every cycle. That means the sampler output is periodic, and therefore the average is always zero regardless of the frequency (or phase) difference [9]. An example of $N = 2.25$ is illustrated in Figure 8.8(b). Because 2.25 f_{ref} is sampled by f_{ref}, the sampled signal is periodic with 0.25 f_{ref} because of aliasing. Reference [8] has proposed to insert a variable delay in the reference path as shown in Figure 8.8(a). The delay is modulated to make the edge-to-edge intervals of the reference clock be multiples of T_{VCO}. That is, for the example of $N = 2.25$, the intervals become like $\{2T_{\mathrm{VCO}}, 2T_{\mathrm{VCO}}, 2T_{\mathrm{VCO}}, 3T_{\mathrm{VCO}}\}$, which is a first-order delta-sigma modulation. As a result, the subsampling is back to synchronous so that it produces a proper non-zero output. Since the fractional spur is highly dependent on the delay resolution and linearity, the variable delay line must be capable of producing fine resolutions and guaranteeing linearity while working with a wide dynamic range, which is set by the VCO operating frequency and the order of the DSM [10]. Such a requirement increases hardware overhead considerably. In addition, an open-loop delay line is sensitive to the PVT variations. In [10], in order to address those issues, a phase interpolator is placed in the feedback path and provides a coarse delay control. The phase interpolator utilizes multiphase generated from the PLL loop hence taking benefit of the PVT-insensitive nature of the loop. Moreover, the coarse/fine

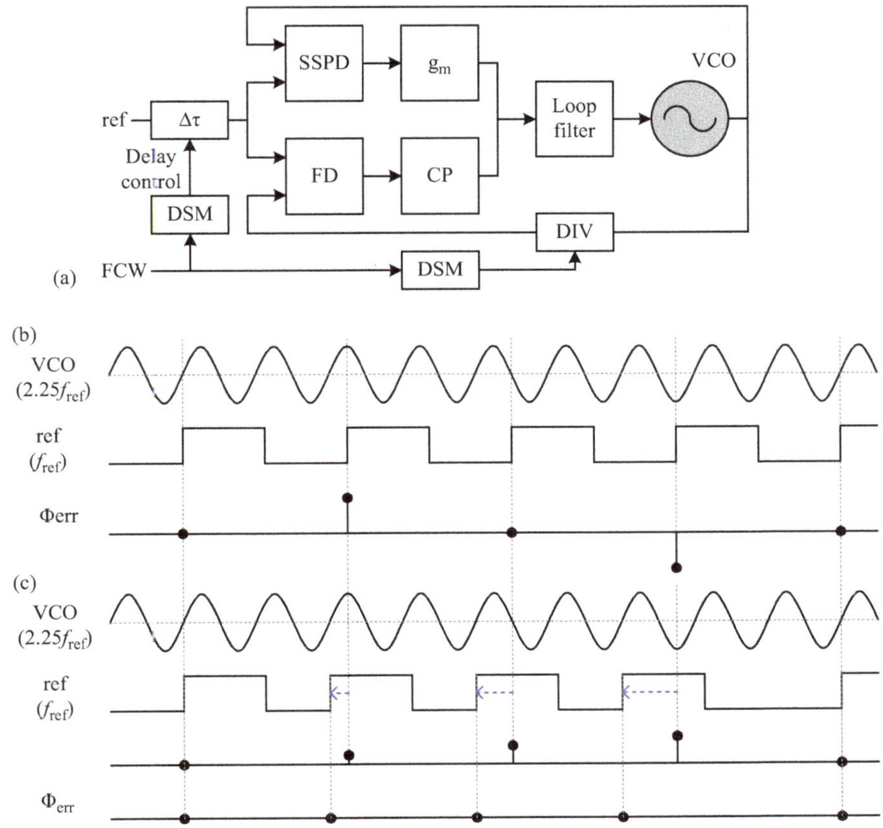

Figure 8.8. Fractional-N SS-PLL: (a) SS-PLL with delay control in reference path, (b) conventional SS-PLL waveforms at harmonic frequency, and (c) SS-PLL with delay control waveforms at harmonic frequency

control reduces the dynamic range of the delay line so the design overhead on capturing both fine resolution and wide range is relaxed.

References

[1] Gao X, Klumperink E, Nauta B. Sub-sampling PLL techniques. In *2015 IEEE Custom Integrated Circuits Conference (CICC)*. San Jose, CA: IEEE; 2015 Sep 28 (pp. 1–8).

[2] Gao X, Klumperink EA, Bohsali M, Nauta B. A low noise sub-sampling PLL in which divider noise is eliminated and PD/CP noise is not multiplied by N_2. *IEEE Journal of Solid-State Circuits*. 2009;44(12):3253–3263.

[3] Sogo K, Toya A, Kikkawa T. A ring-VCO-based sub-sampling PLL CMOS circuit with −119 dBc/Hz phase noise and 0.73 ps jitter. In *2012 Proceedings of the ESSCIRC (ESSCIRC)*. Bordeaux: IEEE; 2012 (pp. 253–256).

[4] Nagam SS, Kinget PR. A low-jitter ring-oscillator phase-locked loop using feedforward noise cancellation with a sub-sampling phase detector. *IEEE Journal of Solid-State Circuits*. 2018;53(3):703–714.

[5] Cho SY, Kim S, Choo MS, *et al.* A 2.5–5.6 GHz subharmonically injection-locked all-digital PLL with dual-edge complementary switched injection. *IEEE Transactions on Circuits and Systems I: Regular Papers*. 2018;65 (9):2691–702.

[6] Nagam SS, Kinget PR. A 0.008 mm 22.4 GHz type-I sub-sampling ring-oscillator-based phase-locked loop with a −239.7 dB FoM and −64dBc reference spurs. In *2018 IEEE Custom Integrated Circuits Conference (CICC)*. San Diego, CA: IEEE; 2018 (pp. 1–4).

[7] Grimaldi L, Bertulessi L, Karman S, *et al.* 16.7 A 30GHz digital sub-sampling fractional-N PLL with 198 fs rms Jitter in 65nm LP CMOS. In *2019 IEEE International Solid-State Circuits Conference-(ISSCC)*. San Francisco, CA: IEEE; 2019 (pp. 268–270).

[8] Chang WS, Huang PC, Lee TC. A fractional-N divider-less phase-locked loop with a subsampling phase detector. *IEEE Journal of Solid-State Circuits*. 2014;49(12):2964–2975.

[9] Bae W. Frequency acquisition technique for injection-locked clock generator using asynchronous-sampling frequency detection. *Electronics Letters*. 2017;53(18):1240–1242.

[10] Narayanan AT, Katsuragi M, Kimura K, *et al.* A fractional-Nsub-sampling PLL using a pipelined phase-interpolator with an FoM of −250 dB. *IEEE Journal of Solid-State Circuits*. 2016;51(7):1630–1640.

Chapter 9
Phase noise suppression techniques 2: all-digital PLL

9.1 Introduction

As CMOS technology scales down, several challenges have been raised which degrades the performance of the analog charge-pump PLLs. For example, the increasing leakage current of the loop filter capacitor degrades the reference spur, the decreasing output impedance of CMOS device increases the pump-current mismatch, and the severe PVT variations make it almost impossible to have the optimum loop bandwidth over the PVT variations. Note that most of the challenges are caused in the analog loop filter. Therefore, the main motivation of the all-digital PLL (ADPLL) is replacing the analog loop filter to the digital loop filter. In a strict sense, the ADPLL refers to a PLL exclusively built from digital function blocks and does not contain any passive component. In a stricter sense, all components of ADPLL are synthesizable. In general, however, a broad sense of ADPLL definition is used such that a PLL consists of digital components (especially the digital loop filter) and digital equivalents. In this chapter, the broad sense of ADPLL will be introduced.

A simplified conversion from the charge-pump PLL to the ADPLL is illustrated in Figure 9.1. The analog loop filter, which consists of the charge-pump and the RC low-pass filter, is replaced by the synthesizable digital loop filter. Therefore, the phase detector and the VCO should be replaced by equivalents that provide digital interfaces. The equivalents of the phase detector and the VCO are time-to-digital converter (TDC) and digitally controlled oscillator (DCO), respectively. On the downside, the digital-friendly nature of the TDC and the DCO introduces new jitter sources to the ADPLL, which we do not have to care in a charge-pump PLL. The TDC produces the digital bits in proportional to the input phase error. Because of the quantization nature of the digital signal, the quantization noise is introduced by the TDC. The DCO converts the digital bits provided from the digital loop filter to the frequency. It would be easier to understand when we imagine a VCO preceded by a digital-to-analog converter (DAC). Unlike the VCO, the output frequency of the DCO is quantized so that the DCO is not able to produce an exactly desired frequency. Therefore, the DCO frequency cycles around the intended frequency at the steady-state, which leads to a deterministic jitter. Such cycling is called a limit-cycle. On the other side, the ADPLL provides a lot of

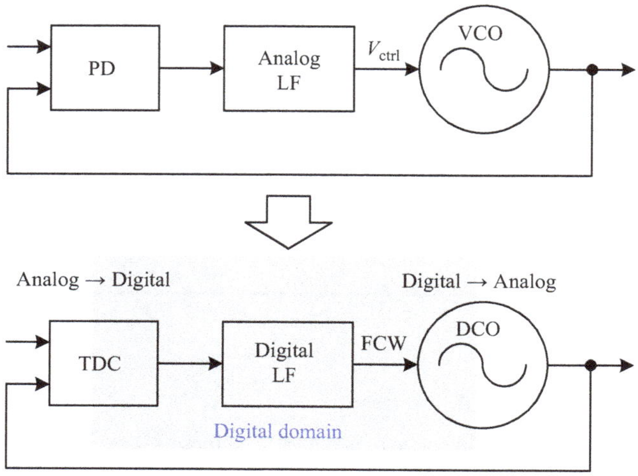

Figure 9.1 ADPLL block diagram

advantages over the charge-pump PLL. At first, besides the issues mentioned above (leakage current, PVT sensitivity, and output impedance), the digital implementation easily benefits from the technology scaling. Designs become more portable across different process technologies and silicon area reduces as the technology shrinks. Moreover, the information gathered from the TDC can be processed more flexibly so that a bit more complex functions (i.e., adaptive loop bandwidth control) are easily integrated into the digital loop filter.

9.2 ADPLL building blocks

9.2.1 Digital loop filter

Whereas the implementation of analog loop filter relies on bulky passive devices (i.e., resistor and capacitor), the digital loop filter (DLF) can be implemented much compact. Figure 9.2 shows DLF examples, which are implemented infinite-impulse-response (IIR) filters in z-domain.

As described in Chapter 8, a PLL loop filter includes a proportional (phase) control and an integral (frequency) control. The proportional and integral terms can be implemented in the DLF as shown in Figure 9.3, where α and β are referred to as integral gain and proportional gain. The transfer gain of the DLF is written as

$$H(z) = \beta + \frac{\alpha}{1 - z^{-1}} = \frac{(\alpha + \beta) - \beta z^{-1}}{1 - z^{-1}} \tag{9.1}$$

Basically, since the digital and the analog loop filter do the same role in the PLL, their transfer function should be compatible with each other [1]. The

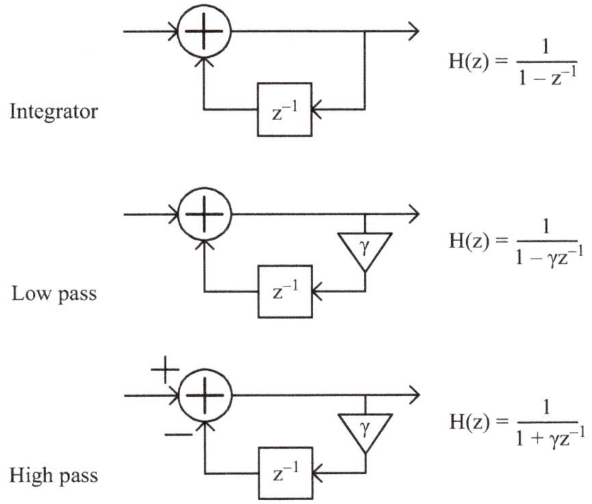

Figure 9.2 DLF examples

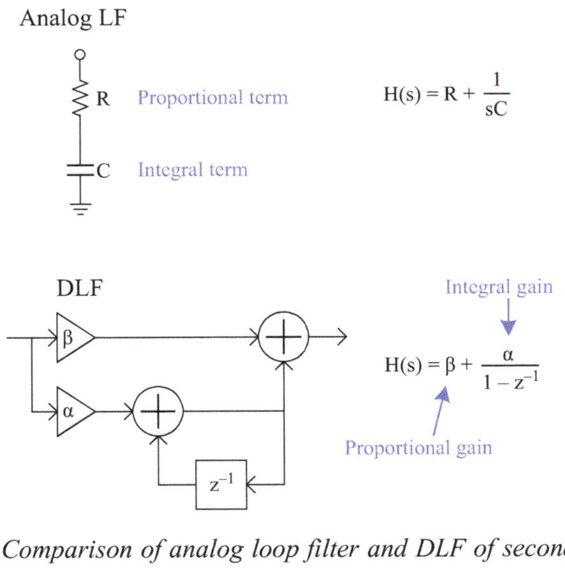

Figure 9.3 Comparison of analog loop filter and DLF of second-order PLL

s-domain analog transfer function is transformed into the digital z-domain by using the bilinear transform (9.2):

$$s = \frac{2}{T_s} \cdot \frac{1 - z^{-1}}{1 + z^{-1}} \tag{9.2}$$

where T_s is the sampling period of the ADPLL, which is generally the period of the reference clock. Applying the bilinear transform to the analog loop filter, the transfer function becomes:

$$H(s) = R + \frac{1}{sC} \rightarrow H(z) = \frac{\left(\frac{T_s}{2C} + R\right) + \left(\frac{T_s}{2C} - R\right)z^{-1}}{1 - z^{-1}} \tag{9.3}$$

By comparing the coefficient of (9.1) and (9.3), we can obtain:

$$\alpha = \frac{T_s}{C} \tag{9.4}$$

$$\beta = R - \frac{T_s}{2C} \tag{9.5}$$

The phase margin is obtained by substituting (9.4) and (9.5) to (6.17) as

$$\mathrm{PM} = \arctan\left(\left(\beta + \frac{T_s}{2C}\right)\frac{T_s}{\alpha}\omega_c\right) \tag{9.6}$$

where we can find the ratio of α and β is constrained by the phase margin as

$$\frac{\beta}{\alpha} = \frac{1}{T_s}\frac{\tan(\mathrm{PM})}{\omega_c} - \frac{1}{2} = \frac{1}{2\pi}\cdot\frac{f_{ref}}{f_c}\cdot\tan(\mathrm{PM}) - \frac{1}{2} \tag{9.7}$$

Equation (9.7) implies that a larger proportional gain leads to a higher phase margin, which we already learned from the charge-pump PLL example.

9.2.2 Time-to-digital converter

TDC is defined as a circuit that converts a time difference between two events into a digital representation. In ADPLL, TDC is used to measure the phase difference between the reference and the feedback clock, for replacing the role of the phase detector. Because of the nature of the quantized digital output, a TDC has a trade-off between the detectable phase range and phase resolution (Figure 9.4). Assuming an N-bit TDC, the resolution is set by dividing the dynamic range by 2^N-1, which implies that the range should be reduced for a finer resolution. Increasing the number of bits can increase the range without sacrificing the resolution, but it also increases the power consumption and the hardware complexity exponentially. Figure 9.5 shows a relation between the resolution and the quantization noise. Due to the quantized step, the digital output deviates from the ideal value. The amount of the deviation is the quantization error, which is a periodic sawtooth function shown in Figure 9.5(b), whose peak-to-peak amplitude equals to the resolution (Δ_{res}). Since the digital output can be expressed as the sum of the ideal value and

---

$$\mathrm{PM} = \arctan(RC\omega_c) = \arctan\left(\frac{\omega_c}{\omega_z}\right) \tag{6.17}$$

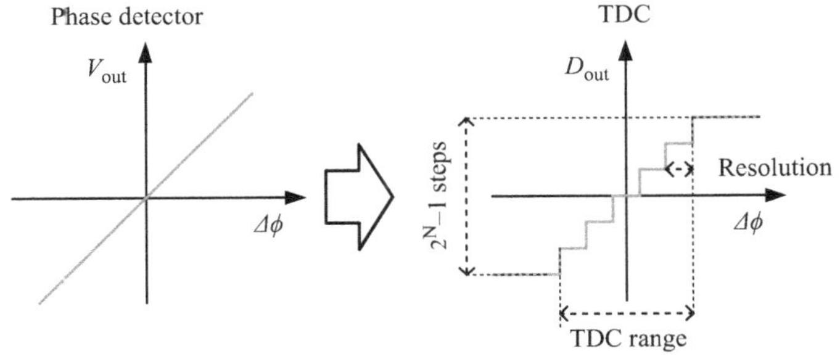

Figure 9.4 Transfer curve of time-to-digital converter

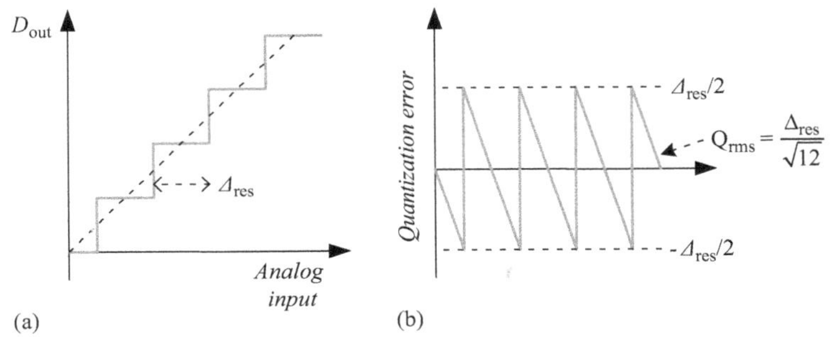

Figure 9.5 Quantization noise

the quantization error, the quantization noise of the digital output is calculated from Figure 9.5(b) as

$$Q_{rms}^2 = \frac{1}{T}\int_0^T \left(\frac{\Delta_{res}}{2}\cdot\frac{x}{T}\right)^2 dx = \frac{\Delta_{res}^2}{12} \tag{9.8}$$

The detectable range is also important to be able to react to a large phase error and achieve a fast locking [2]. In addition to the dynamic range, resolution, and power consumption, linearity metrics such as differential nonlinearity (DNL) and integral nonlinearity (INL) are also important for the TDC. For an ADC, a difference between analog voltages that correspond to consecutive digital codes is ideally one least-significant bit (LSB). However, in practice, the difference deviates from the ideal one LSB, and the DNL is defined as the deviation over the one LSB. On the other hand, the INL is defined as the deviation of the actual ADC curve to the straight line that connects the start point and the end point. A DNL and INL example is provided in Figure 9.6.

From now on, let us study a few examples of TDC implementation. A delay-chain TDC implementation shown in Figure 9.7 is the most primitive one. Basically, it is similar to a chain of bang-bang phase detectors, which takes two clock signals,

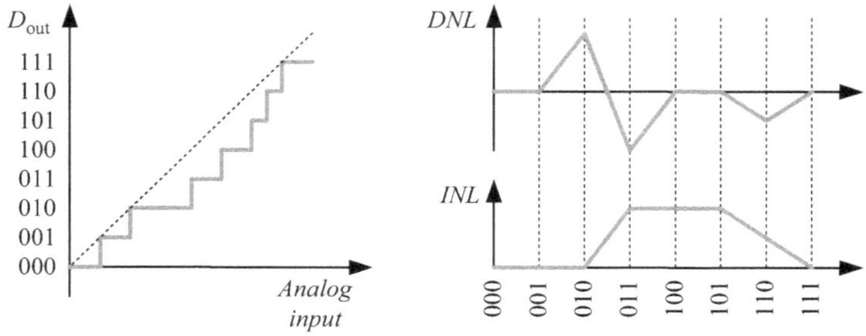

Figure 9.6 INL and DNL

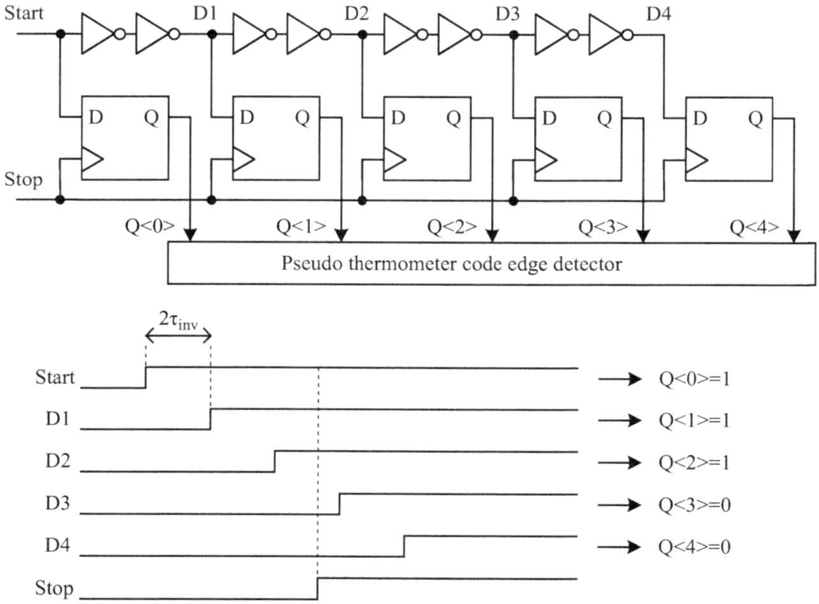

Figure 9.7 Delay-chain TDC with two-inverter delay resolution

Start and *Stop*, as input. Each of the bang-bang phase detectors compares the *Stop* signal to the gradually delayed *Start* signal. For the example shown in Figure 9.6, the output of the first three D flip-flops (Q<2:0>) becomes high but the last two (Q<4:3>) becomes low; because of the delay elements the *Start* starts lagging from D3. Note that the D flip-flops can also take the *Start* signal for the clock and the *Stop* signal for the input, then the polarity of the output is flipped (i.e., Q<4:0>: 11100 → 00011). Here, the delay element is implemented with two-stage inverters not to flip the polarity of the input clock. Therefore, the time resolution of this TDC is given to the twice of inverter delay, which is generally a few tens of picoseconds.

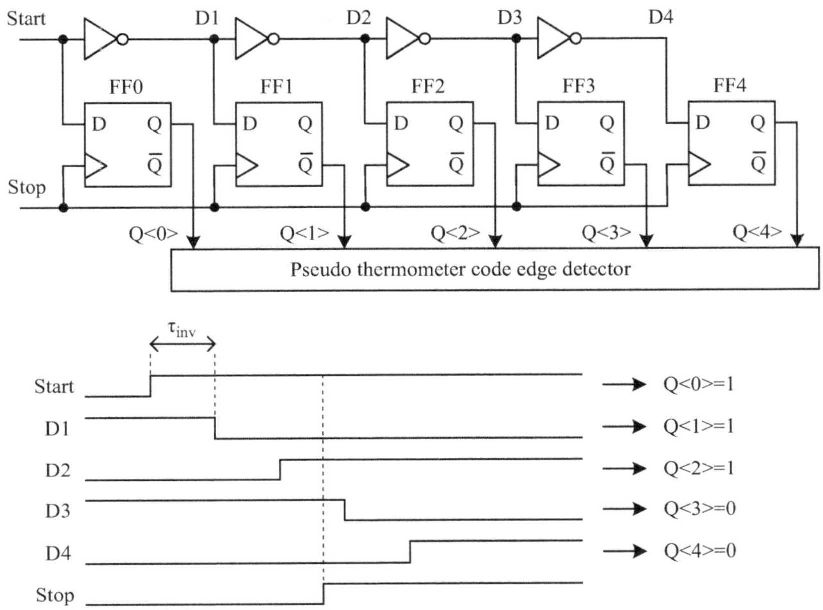

Figure 9.8 Delay-chain TDC with one-inverter delay resolution

From (9.8), it is equivalent to adding a substantial amount of RMS jitter to the output of the TDC.

The time resolution is reduced to one-inverter delay when we use the QB of the D flip-flops in an odd order, as shown in Figure 9.8. The flipped polarity of the input clock for the odd flip-flops is simply corrected by inverting the flip-flop output. However, a practical flip-flop has a different setup time for input data "1" and "0," which introduces a mismatch between the even and odd paths. Moreover, the rising time and falling time (delays) are not balanced in a practical inverter as well. The mismatch introduces a periodic fluctuation on DNL and INL of the TDC, which degrades the effective resolution.

In [3], a pseudo-differential configuration with symmetric D flip-flops has been proposed to realize one-inverter-delay resolution while resolving the even-odd mismatch (Figure 9.9). A differential sense-amplifier-based D flip-flop is used, where the setup time dependency on the input polarity is eliminated due to the differential nature [4]. Pseudo-differential inverter is used as a delay element for the differential input clock. Since two sides of the pseudo-differential inverter have different polarities, the effect of unequal rising and falling time is averaged out by the symmetric D flip-flop.

Although the pseudo-differential TDC enables reducing the resolution by half, an inverter delay is still too large to suppress the quantization jitter. A Vernier TDC is proposed to obtain a finer resolution less than one-inverter delay (Figure 9.10) [5]. The basic concept of the Vernier TDC is to introduce a delay chain to the *Stop* path

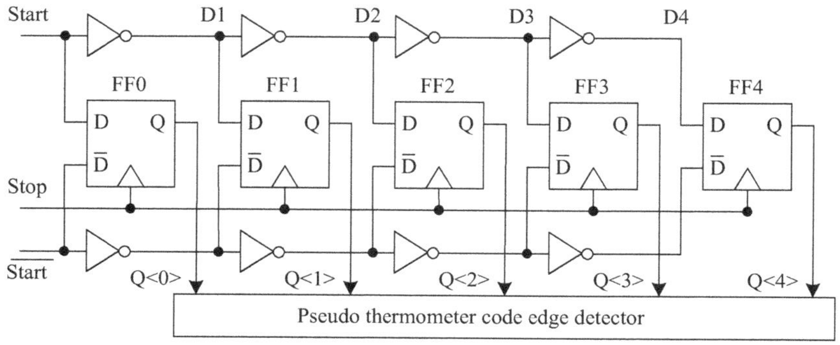

Figure 9.9 Delay-chain TDC with one-inverter-delay resolution without even-odd mismatch

as well so that the time resolution becomes the delay difference between the *Start* path and the *Stop* path. One of the most important constraints here is the *Start* path should be always slower than the *Stop* path. If one of the delay elements in the faster path becomes slower due to random mismatch, the TDC loses the monotonicity. That means the achievable resolution is constrained by the mismatch between the delay elements. Therefore, a larger delay cell, which leads to a higher power consumption, is required to achieve a finer resolution.

The previous examples of TDC suffer from the range-resolution trade-off, which becomes more critical for a wide-range PLL. For a wide range and a fine resolution, too many delay cells and D flip-flops are needed, which result in large area and power consumption. The following three examples of two-step TDC (Figure 9.11), logarithmic TDC (Figure 9.12), and ring TDC (Figure 9.13) show advanced techniques that try to break the trade-off. Reference [6] proposed a two-step TDC that incorporates both the delay chain TDC and Vernier TDC. The first stage is similar to the typical delay-chain TDC which takes the Start signal as the clock input for the D flip-flop, but an additional D flip-flop stage is placed at the front of the TDC. In the additional stage, the data input and the clock input are opposite to those in the TDC stages, and therefore the output is synchronized to the *Stop*. The delay-chain TDC is followed by AND gate array. Since the AND output is "0" when the adjacent TDC bits are the same but becomes "1" when the bits are different, the AND array detects the transition of the thermometer code. The AND array outputs are summed through the OR gate, that is, the OR output (FCLK2) is triggered from "0" to "1" by the thermometer transition. As a result, the FCLK2 is synchronized to the delayed *Start* edge where the thermometer transition happens. The AND OR gates are also used for the output of the additional stage (FCLK1) in order to equalize the delay of FCLK1 and FCLK2. The FCLK1 is synchronized to the *Stop* edge while the FCLK2 is synchronized to the delayed *Start* edge right after the *Stop* edge; the time difference of FCLK1 and FCLK2 equals to the residue of the first stage TDC, which is fed to the second Vernier stage. Since the

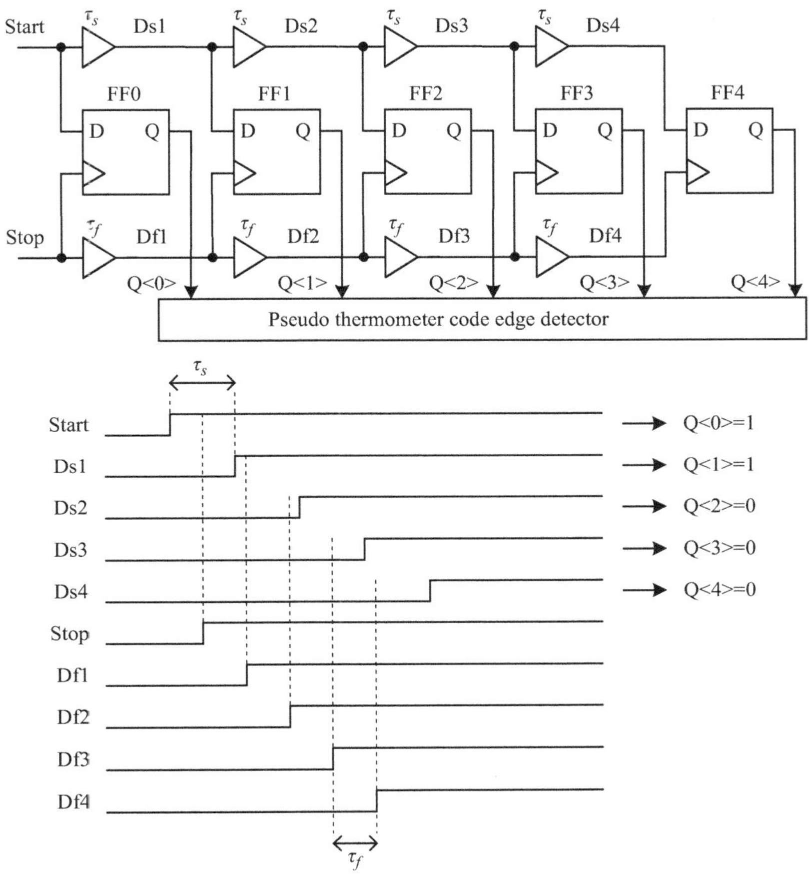

Figure 9.10 Vernier TDC

maximum residue is τ_{d1}, the required range of Vernier TDC is only τ_{d1}. In such two-step TDC architecture, it is important to calibrate the resolution ratio between the first stage and the second stage to normalize the output bits from the two stages.

The logarithmic TDC incorporates an exponentially increased delay chain rather than a uniform delay chain, as shown in Figure 9.12 [7]. For example, if the delay increases by a factor of two through the chain, the TDC does a binary search to convert the time difference to the digital bits, Q<n:0>. Therefore, with N flip-flops, the detectable range of TDC is ideally $(2^N-1)\tau_{d1}$, which is much wider than $N\tau_{d1}$ of the conventional TDC. However, due to the limitation of the practical implementation, the exponentially increased delay chain suffers a severe mismatch between stages compared to a uniform delay chain, which leads to a degradation of the linearity. Therefore, it should be followed by an additional linearization logic, for example, a look-up-table (LUT) [7].

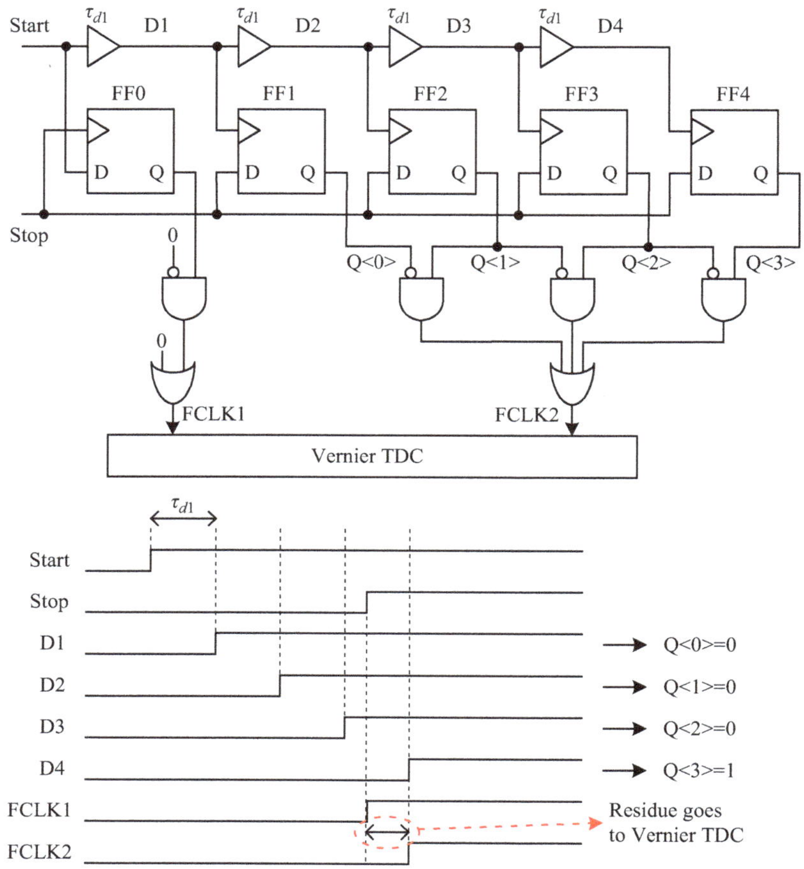

Figure 9.11 Two-step TDC

Figure 9.13 shows a simplified circuit diagram of a ring TDC. Instead of adding delay chain and D flip-flops, the ring TDC reuses the delay units by forming a ring oscillator once the *Start* signal goes high [8]. As a result, the clock edge triggered by the *Start* propagates through the ring until the *Start* returns to low, and the flip-flop array continuously compares the *Stop* and the delayed *Start* while the delay accumulatively increases over cycles. A counter takes the last phase of the oscillator clock and counts number of cycles until the rising edge of the *Stop* is triggered, which can be regarded as coarse bits. From the fact that the ring is formed when the Start is high, the detection range is ideally increased to half of the reference period. Moreover, because the ring TDC reuses the delay units over cycles, it has less mismatch issues between the delay units hence exhibiting better linearity, compared to the conventional delay-chain TDC. The main drawback of the ring TDC is power consumption, since the clock edge propagates continuously

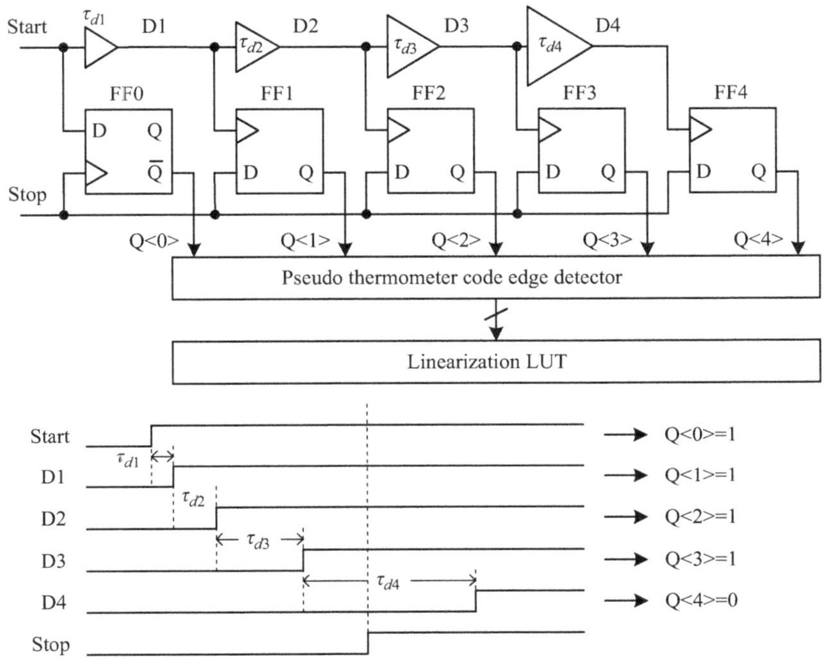

Figure 9.12 Logarithmic TDC

through the ring oscillator, hence dissipating a static power. A Vernier ring TDC which improves the resolution is implemented by forming a second ring at the *Stop* delay path and adding another counter that counts the second oscillator clock [9], at the cost of a reduced detection range and increased power consumption.

Earlier in the text, we studied various examples of TDC implementation. One of the common points we can find is that the resolution (Δ_{res}) of all of them relies on CMOS delay elements (i.e., CMOS inverter), which are very sensitive to PVT variations. Note that we can define gain of TDC (K_{TDC}) as a slope of TDC input-output transfer curve, which equals Δ_{res}, neglecting any mismatch source between delay units. In modern CMOS technologies, the fastest PVT corner is 2–3× faster than the slowest PVT corner, which means the transfer function of an ADPLL becomes no more insensitive to the PVT variations due to the K_{TDC}. Recalling from the Introduction of this chapter, the sensitivity to the PVT variation of analog PLLs is one of the primary motivations of ADPLL. Note that a TDC dissipates much more power compared to a conventional phase detector. Thus, one can say that if a TDC makes an ADPLL be PVT-sensitive, the ADPLL is no longer very attractive. In order to address this challenge, a DLL-based PVT stabilization technique has been proposed in [10]. The DLL uses a replica of the delay chain of the TDC and keeps the delay of the replica constant regardless of the PVT variations. The delay chain of the TDC shares the control voltage set by the DLL. As a result, the TDC

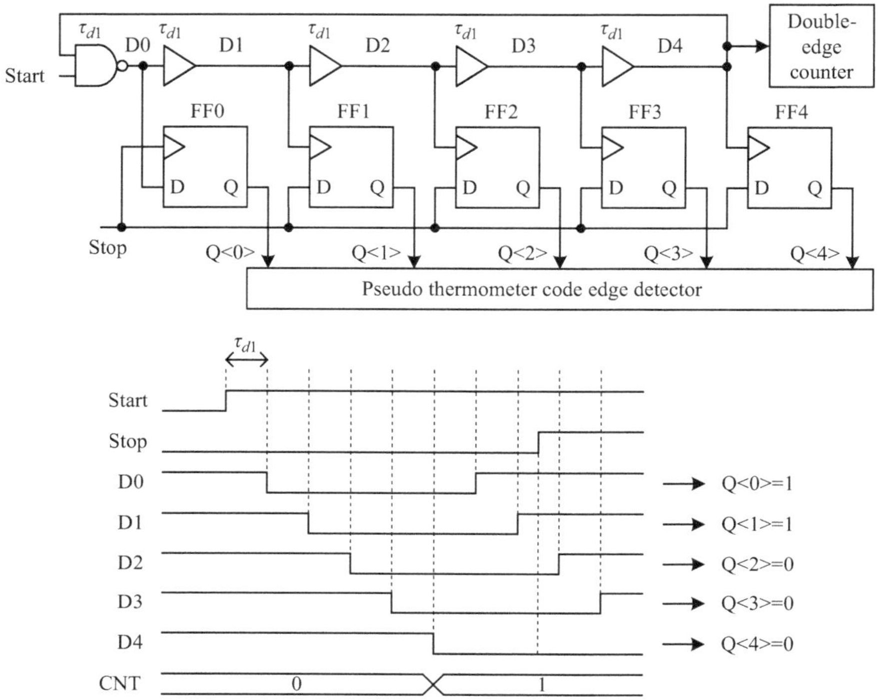

Figure 9.13 Ring TDC

resolution is set by the DLL hence it is insensitive to the PVT variations. Note that the TDC resolution equals to

$$\Delta_{\text{res}} = \tau_d = \frac{T_{\text{CLK}}}{N} \tag{9.9}$$

where T_{CLK} and N are the DLL clock period and the number of stages of the replica delay chain. The DLL-based approach can also be adopted for Vernier TDC, by using two DLLs, a slow DLL for the *Start* path and a fast DLL for the *Stop* path as shown in Figure 9.14 [10]. Additional delay stages are introduced to the replicas chain in the fast DLL, so the delay per stage in the fast DLL becomes shorter than that in the slow DLL. As a result, the resolution of the Vernier TDC becomes:

$$\Delta_{\text{res}} = \tau_s - \tau_f = \frac{T_{\text{CLK}}}{N} - \frac{T_{\text{CLK}}}{N+m} = \frac{m \cdot T_{\text{CLK}}}{N(N+m)} \tag{9.10}$$

where m is the number of stages added to the fast DLL. In the same way, a PLL-based stabilization can be introduced for ring TDCs [11].

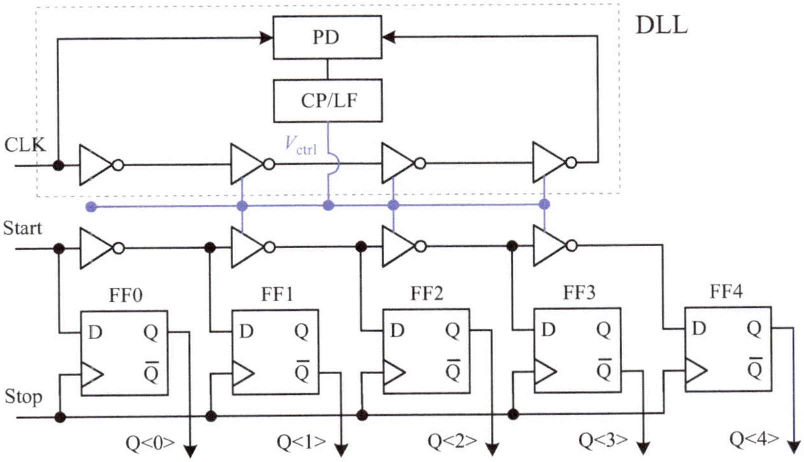

Figure 9.14 DLL-stabilized TDC

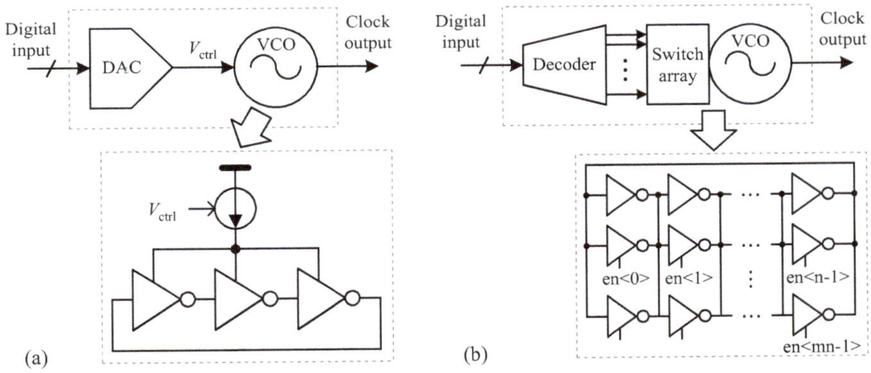

Figure 9.15 Example of DCO implementation methods: (a) explicit DAC + VCO and (b) embedded DAC

9.2.3 Digitally controlled oscillator

DCO is the most critical component in ADPLL implementation since it is one of the dominant sources of the quantization noise as well as the phase noise. A DCO converts digital input to frequency, which is analog, so its underlying functionality is almost analog similar to typical ADCs or DACs. In fact, a DCO is a kind of (or another name of) digital-to-frequency convertor. DCO implementations can be classified into two categories. A straightforward implementation is to divide such digital-to-frequency conversion into explicit two steps: digital-to-analog voltage conversion and voltage-to-frequency conversion, which is shown in Figure 9.15(a) [12,13]. At the front of a VCO which is something similar used in analog PLLs (i.e., ring-VCO and LC-VCO), a DAC converts digital input from DLF to the analog voltage for the VCO. Because there is an explicit analog control voltage like an analog PLL, this implementation can be regarded as an analog-like approach. Because of its analog nature, it shares design challenges with analog PLLs, such as the reduced voltage headroom issue. On the other hand, the analog voltage is replaced by using an embedded DAC instead of an explicit DAC. For example, if we replace the capacitance of LC-tank oscillator to a capacitor DAC, the analog voltage is removed. For a ring oscillator, the most straightforward way is to control the number of stages. We can also control the number of drivers for each delay stage as shown in Figure 9.15(b), which is a kind of resistive tuning [14]. Because the resolution achieved from the stage selection is limited, it is generally combined with the resistive tuning for wide range and fine resolution [15]. As an alternative way, a capacitive tuning is enabled by placing a capacitor DAC at the output of each delay stage. In general, the resistive tuning has a wider dynamic range, but the capacitive tuning provides a finer resolution. The combination of the coarse-resistive tuning and the fine-capacitive tuning can also be used, at the cost of reduced speed due to the additional nodal capacitance.

Because a DCO has the digital input, a non-ideal switching at the input may introduce an additive noise on the DCO phase noise. For example, a digital glitch which is introduced by a nonuniform timing between digital bits leads to a significant phase error. Unlike analog PLL, the output frequency of ADPLL is quantized so that an ADPLL cannot lock to an exact frequency even though we neglect any nonideal conditions (i.e., noise). Instead, an ADPLL achieves lock by dithering its frequency around the intended frequency, so that the average frequency equals the intended frequency. In other words, at steady-state of ADPLL, the DCO input dithers between two adjacent codes where the intended frequency locates somewhere between them. This kind of cycling behavior is called a limit cycle. In this ideal case the DCO input dithers between the adjacent codes and the maximum frequency discrepancy is the frequency resolution of the DCO. However, in practice, each digital bit has a different delay from the source, so their timing cannot be identical. In case of switching multiple bits at the same time, this timing mismatch introduces a switching noise on the digital code, which is called a glitch. Even though the digital code moves by 1 LSB amount, multiple bits can be switched, for example, switching of 011 and 100 in binary-weighted code. Figure 9.16 illustrates the glitch issue of 3-bit, binary-weighted DCO. If the digital input dithers between 011 and 100 at the steady-state, all the bits should switch simultaneously (0=>1, 1=>0, 1=>0). However, in practice, there is no literal "simultaneous switching," so here we can assume that the LSB switches earlier the MSB (for a typical DCO implementation, MSB has a larger capacitive loading). Then the 011-to-100 switching actually becomes a sequential switching of 011-010-000-100. We can easily observe that the DCO frequency hits the minimum while switching. At the same manner, the DCO also hits the maximum frequency while dithering so that the maximum frequency discrepancy becomes the entire frequency range of the DCO. Since the amount of phase dithering is the integration of the DCO

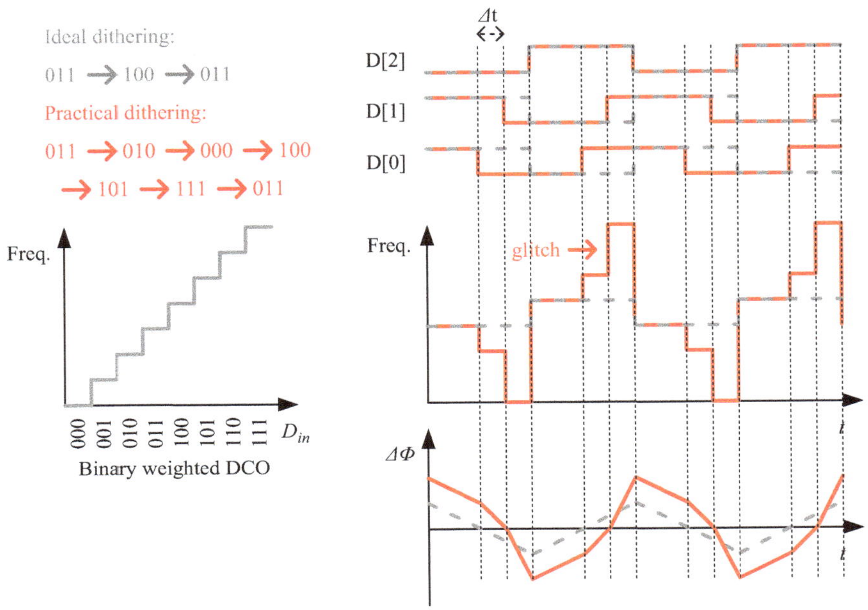

Figure 9.16 Glitch in a binary-weighted DCO

frequency over time, the glitch introduces a substantial amount of periodic phase error as shown in Figure 9.16 over the ideal case.

To prevent such glitch, thermometer code, where only one-bit state is flipped during adding (or subtracting) by one LSB, is frequently used in DCO implementation. However, the thermometer code is too lengthy (i.e., 10-bit binary code is equivalent to 1024-bit thermometer code), hence the decoder and signal routing become inefficient. Two-dimensional decoding shown in Figure 9.17 is widely used because of the reduction of bit length. For example, once divided into row and column, 1024-bit thermometer code is reduced to 32-bit row code and 32-bit column code. Because the row code and the column code serve as, respectively, coarse and fine-tuning, the glitch issue should be considered at the boundary of the row code switching. Basically, a DAC element is turned on when both of row and column codes are high, but it is turned on regardless of row and column codes when the next row code is high, as shown in Figure 9.17(a), if we use identical cells for both even and odd rows. Note that the last column bit is fixed to low, which makes the state of the last cell solely depends on the next row code. As a result, no column code is flipped while the next row code switches low-to-high. However, multiple column bits should be switched once the next row starts turning on, which leads to a glitch as long as the cell turns on when the column code is high. Recalling from the glitch issue in binary-weighted DCO, only one of the control bits should be flipped at the same time. Therefore, using two different cells for even and odd rows is a

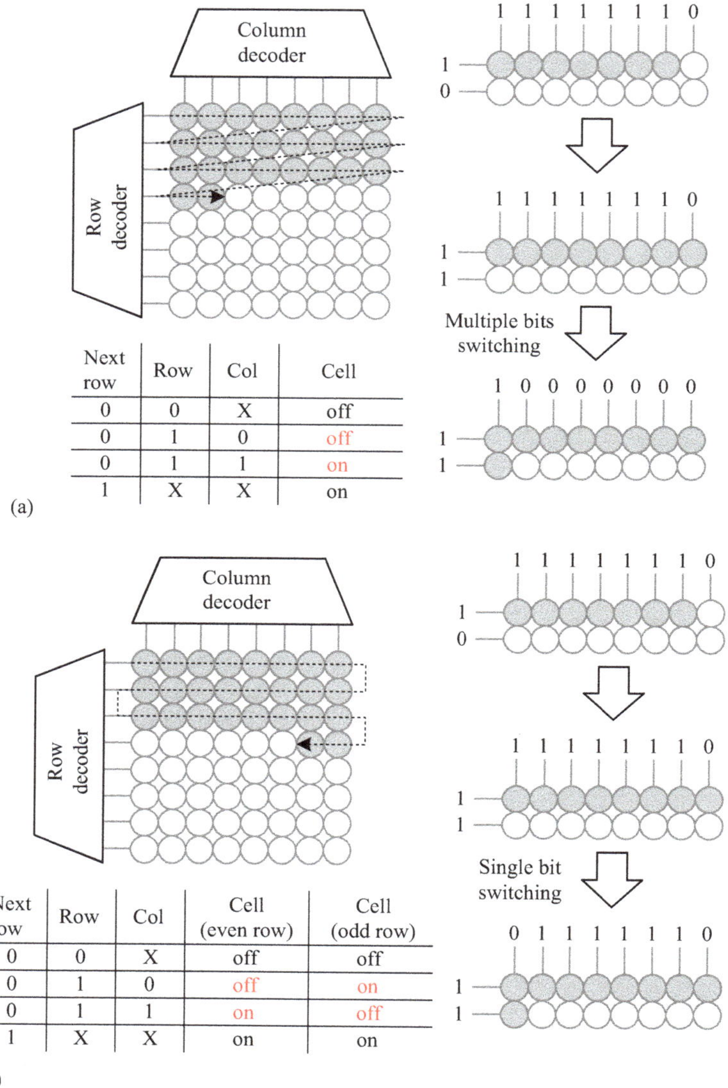

(a)

Next row	Row	Col	Cell
0	0	X	off
0	1	0	off
0	1	1	on
1	X	X	on

(b)

Next row	Row	Col	Cell (even row)	Cell (odd row)
0	0	X	off	off
0	1	0	off	on
0	1	1	on	off
1	X	X	on	on

Figure 9.17 (a) Two-dimensional thermometer-coded DAC and (b) glitch-free implementation

better implementation to eliminate glitches, which is shown in Figure 9.17(b) [13]. For the cells in the odd row, they are turned on when the column code is low while the dependency on row codes remains the same as the even row. In this glitch-free implementation, when the DCO control code increases, the column decoder lets the thermometer code increase if the last active row is even but decreases if the last row is odd. As a result, only one-bit state changes at any of the switching boundary hence the glitch is eliminated.

9.3 Quantization noise and jitter

9.3.1 Linearized model of ADPLL

In Chapter 6, we studied the jitter transfer function of noise sources and trade-offs, and how they affect the PLL jitter performance. In ADPLL, there are two analog-to-digital (or digital-to-analog) interfaces where the quantization noises are introduced. This chapter will cover the jitter transfer functions of the quantization noises. Figure 9.18 shows a linearized model of an ADPLL. The TDC is modeled as a combination of a phase-to-time converter and a phase-to-digital convertor. Assuming that ϕ_{in} is the input phase error of the TDC, the time error detected by the phase-to-time converter is expressed as

$$t_{P2T} = \frac{T_{\text{REF}}}{2\pi}\phi_{\text{err}} \tag{9.11}$$

where T_{REF} is the reference clock period. The time error is quantized by the TDC resolution (Δ_{TDC}) during the time-to-digital conversion, so the digital output of the TDC is written as

$$D_{\text{TDC}} = \frac{T_{\text{REF}}}{2\pi}\cdot\frac{1}{\Delta_{\text{TDC}}}\cdot\phi_{\text{err}} \tag{9.12}$$

which leads to the TDC gain expression:

$$G_{\text{TDC}} = \frac{D_{\text{TDC}}}{\phi_{\text{err}}} = \frac{T_{\text{REF}}}{2\pi}\cdot\frac{1}{\Delta_{\text{TDC}}} \tag{9.13}$$

The TDC quantization noise is modeled to be added at the output of the phase-to-time conversion. On the other hand, we can express the DCO gain just by replacing K_{VCO} to K_{DCO} ($2\pi K_{\text{DCO}}/s$), once we define the K_{DCO} as the frequency

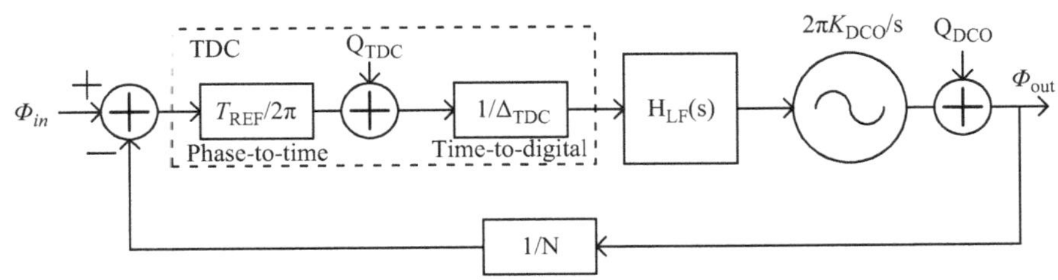

Figure 9.18 Linearized s-domain model of ADPLL

step per LSB (Hz/bit). The DCO quantization noise is also modeled to be added at the output of the DCO. The open-loop gain of an ADPLL is written as

$$T(s) = \frac{T_{\mathrm{REF}}}{2\pi} \cdot \frac{1}{\Delta_{\mathrm{TDC}}} \cdot H_{\mathrm{LF}}(s) \cdot \frac{2\pi K_{\mathrm{DCO}}}{s} = \frac{T_{\mathrm{REF}} K_{\mathrm{DCO}} H_{\mathrm{LF}}(s)}{\Delta_{\mathrm{TDC}} s} \tag{9.14}$$

We can easily achieve (9.14) by substituting I_{CP}, K_{VCO}, and $Z_{\mathrm{LF}}(\mathrm{s})$ of a charge-pump PLL to $T_{\mathrm{REF}}/\ \Delta_{\mathrm{TDC}}$, K_{DCO}, and $H_{\mathrm{LF}}(\mathrm{s})$, respectively [16]. Combining (6.15), (9.4), and (9.5), the cut-off frequency is obtained as

$$\omega_c = \sqrt{2}\frac{T_{\mathrm{REF}}}{\Delta_{\mathrm{TDC}}}\left(\beta + \frac{\alpha}{2}\right)K_{\mathrm{DCO}} \tag{9.15}$$

9.3.2 Quantization noise of TDC

Before deriving the jitter transfer function, the TDC quantization noise is assumed spectrally white, that is, distributed uniformly over frequency. Then, from (9.8), we can obtain the PSD of the quantization noise as

$$\mathrm{S}_{\phi,\mathrm{TDC}}(f) = \frac{Q_{\mathrm{rms}}^2}{f_{\mathrm{REF}}} = \frac{\Delta_{\mathrm{TDC}}^2}{12} \cdot \frac{1}{f_{\mathrm{REF}}} \tag{9.16}$$

where all the frequency components are folded into the range below f_{REF}, which is the sampling frequency of the TDC. Recalling Section 6.2.3, we can obtain the closed-loop transfer function of the TDC quantization noise as

$$H(s) = \frac{T(s)}{1 + T(s)/N} \cdot \frac{2\pi}{T_{\mathrm{REF}}} \tag{9.17}$$

Combining (9.16) and (9.17), the output phase noise by the TDC quantization noise is expressed as

$$\mathrm{S}_{\phi_{\mathrm{OUT,TDC}}}(f) = \mathrm{S}_{\phi,\mathrm{TDC}}(\mathrm{f})|H(s)|^2 = \Delta_{\mathrm{TDC}}^2 \cdot \frac{(2\pi)^2}{12 T_{\mathrm{REF}}} \cdot \left(\frac{T(s)}{1 + \frac{T(s)}{N}}\right)^2 \tag{9.18}$$

where we can find that the TDC quantization noise is low-pass filtered by the loop. The in-band phase noise is simplified to

$$\mathrm{S}_{\phi_{\mathrm{OUT,TDC}}}(f \ll f_c) = \Delta_{\mathrm{TDC}}^2 \cdot \frac{(2\pi)^2}{12 T_{\mathrm{REF}}} \cdot N^2 \tag{9.19}$$

where we can find that the phase noise induced by the TDC is inversely proportional to the square of Δ_{TDC} and N.

$$\omega_c \cong \frac{I_{\mathrm{CP}} R K_{\mathrm{VCO}}}{2N} \tag{6.15}$$

9.3.3 Quantization noise of DCO

The quantization noise of DCO is assumed to be an additive random variable with white spectral density. At the same manner of (9.8), DCO quantization frequency error is expressed as

$$\Delta f_{\text{rms}}^2 = \frac{\Delta f_{\text{res}}^2}{12} \tag{9.20}$$

where Δf_{res} is the frequency resolution of the DCO. Assuming that the quantization noise is distributed equally from DC to the Nyquist frequency of the DCO output ($f_{\text{DCO}}/2$), the single-sideband spectral density of the quantization frequency error becomes:

$$S_{\Delta\text{f,SSB}}(f) = \frac{\Delta f_{\text{res}}^2}{12} \cdot \frac{1}{f_{\text{DCO}}} = \frac{\Delta f_{\text{res}}^2}{12} \cdot \frac{1}{N f_{\text{REF}}} \tag{9.21}$$

Multiplied by the frequency-to-phase transfer function of $2\pi/\text{s}$, the PSD at the DCO output by the quantization noise is obtained as

$$\text{S}_{\phi}(f) = \frac{\Delta f_{\text{res}}^2}{12} \cdot \frac{1}{f_{\text{DCO}}} \cdot \left(\frac{1}{f}\right)^2 \tag{9.22}$$

Because the DCO input is updated at the DLF clock rate (i.e., f_{ref}) but holds at the same value during the cycle, it is a zero-order hold system so that the zero-order hold operation, whose transfer function is the sinc function, should be included [17]. In this chapter, however, we neglect the zero-order hold operation for simplicity. Combined with (6.30), the output phase noise contribution of the DCO quantization noise is obtained as

$$S_{\phi_{\text{OUT,DCO}}}(f) = \text{S}_{\phi,\text{DCO}}(\text{f})|H(s)|^2 = \frac{1}{12} \cdot \frac{1}{f_{\text{DCO}}} \cdot \left(\frac{\Delta f_{\text{res}}}{f}\right)^2 \cdot \left(\frac{1}{1 + \frac{T(f)}{N}}\right)^2 \tag{9.23}$$

In general, the Δf_{res} is proportional to the f_{DCO}, so (9.23) implies that the DCO quantization noise is linearly proportional to the Δf_{res}.

References

[1] Kratyuk V, Hanumolu PK, Moon UK, Mayaram K. A design procedure for all-digital phase-locked loops based on a charge-pump phase-locked-loop analogy. *IEEE Transactions on Circuits and Systems II: Express Briefs*. 2007;54(3):247–251.

$$\frac{\Delta\phi_{\text{out}}}{\Delta\phi_{\text{VCO}}} = \frac{1}{1 + \frac{T(s)}{N}} \tag{6.30}$$

[2] Yu J, Dai FF, Jaeger RC. A 12-bit Vernier ring time-to-digital converter in 0.13 μm CMOS technology. *IEEE Journal of Solid-State Circuits*. 2010;45 (4):830–842.

[3] Staszewski RB, Vemulapalli S, Vallur P, Wallberg J, Balsara PT. 1.3 V 20 ps time-to-digital converter for frequency synthesis in 90-nm CMOS. *IEEE Transactions on Circuits and Systems II: Express Briefs*. 2006;53(3): 220–224.

[4] Nikolic B, Oklobdzija VG, Stojanovic V, Jia W, Chiu JK, Leung MM. Improved sense-amplifier-based flip-flop: Design and measurements. *IEEE Journal of Solid-State Circuits*. 2000;35(6):876–884.

[5] Dudek P, Szczepanski S, Hatfield JV. A high-resolution CMOS time-to-digital converter utilizing a Vernier delay line. *IEEE Journal of Solid-State Circuits*. 2000;35(2):240–247.

[6] Tokairin T, Okada M, Kitsunezuka M, Maeda T, Fukaishi M. A 2.1-to-2.8 GHz all-digital frequency synthesizer with a time-windowed TDC. In *2010 IEEE International Solid-State Circuits Conference Digest of Technical Papers (ISSCC)*. San Francisco, CA: IEEE; 2010 (pp. 470–471).

[7] Lin J, Haroun B, Foo T, *et al.* A PVT tolerant 0.18 MHz to 600MHz self-calibrated digital PLL in 90nm CMOS process. In *2004 IEEE International Solid-State Circuits Conference, 2004. Digest of Technical Papers (ISSCC)*. San Francisco, CA: IEEE; 2004 (pp. 488–541).

[8] Chang HH, Wang PY, Zhan JH, Hsieh BY. A fractional spur-free ADPLL with loop-gain calibration and phase-noise cancellation for GSM/GPRS/EDGE. In *IEEE International Solid-State Circuits Conference, 2008 (ISSCC 2008) Digest of Technical Papers*. San Francisco, CA: IEEE; 2008 (pp. 200–606).

[9] Yu J, Dai FF, Jaeger RC. A 12-bit Vernier ring time-to-digital converter in 0.13 μm CMOS technology. *IEEE Journal of Solid-State Circuits*. 2010; 45(4):830–842.

[10] Hwang CS, Chen P, Tsao HW. A high-precision time-to-digital converter using a two-level conversion scheme. *IEEE Transactions on Nuclear Science*. 2004;51(4):1349–1352.

[11] Chen P, Chen CC, Zheng JC, Shen YS. A PVT insensitive Vernier-based time-to-digital converter with extended input range and high accuracy. *IEEE Transactions on Nuclear Science*. 2007;54(2):294–302.

[12] Kratyuk V, Hanumolu PK, Ok K, Moon UK, Mayaram K. A digital PLL with a stochastic time-to-digital converter. *IEEE Transactions on Circuits and Systems I: Regular Papers*. 2009;56(8):1612–621.

[13] Oh DH, Kim DS, Kim S, Jeong DK, Kim W. A 2.8 Gb/s all-digital CDR with a 10b monotonic DCO. In *IEEE International Solid-State Circuits Conference, 2007. ISSCC 2007. Digest of Technical Papers*. San Francisco, CA: IEEE; 2007 (pp. 222–598).

[14] Olsson T, Nilsson P. A digitally controlled PLL for SoC applications. *IEEE Journal of Solid-State Circuits*. 2004;39(5):751–760.

[15] Hsu TY, Wang CC, Lee CY. Design and analysis of a portable high-speed clock generator. *IEEE Transactions on Circuits and Systems II: Analog and Digital Signal Processing*. 2001;48(4):367–375.

[16] Kratyuk V, Hanumolu PK, Moon UK, Mayaram K. A design procedure for all-digital phase-locked loops based on a charge-pump phase-locked-loop analogy. *IEEE Transactions on Circuits and Systems II: Express Briefs*. 2007;54(3):247–251.

[17] Staszewski RB, Hung CM, Barton N, Lee MC, Leipold D. A digitally controlled oscillator in a 90 nm digital CMOS process for mobile phones. *IEEE Journal of Solid-State Circuits*. 2005;40(11):2203–2211.

Chapter 10

Phase noise suppression techniques 3: injection locking

10.1 Injection locking basics

Injection locking refers to a phenomenon that a periodic charge injection to an oscillator which leads to a frequency shift to the oscillator's free-running frequency. In specific, if the frequency of the injection signal is at the vicinity of the free-running frequency ($f_{free} + \Delta f$), the oscillation frequency becomes the injected frequency instead of the free-running frequency. Figure 10.1 illustrates how the periodic injection shifts the oscillation frequency. It is assumed that there is a small offset (ΔT) between the periods of the free-running oscillation and the injection pulse train. Recalling from (4.23), each pulse injection leads to the phase shift whose amount is proportional to the amount of the injected charges. That means, if the amount of the charge is large enough, the phase shift introduced by the injection is able to lag the next zero-crossing by ΔT. With the periodic pulse train, this period lag happens at every cycle, and therefore the period of the oscillation becomes $T_{free} + \Delta T$, instead of T_{free}.

From (4.23), the maximum shift is calculated as

$$\Delta T_{max} = \frac{T_{free}}{2\pi} |\Gamma(\omega_0 \tau)|_{max} \frac{\Delta q}{q_{max}} \tag{10.1}$$

For the case of injection-locked oscillator (ILO), we can define $\frac{\Delta q}{q_{max}}$ as the injection strength, whose range is between 0 and 1. Assuming sinusoidal waveform (i.e., $V_{swing}\cos\omega_0 t$) and the ISF of (4.51), and substituting CV_{swing} to q_{max}, (10.1) can be simplified to

$$\Delta T_{max} = \frac{T_{free}}{2\pi} \frac{\Delta q}{CV_{swing}} = \frac{T_{free}}{2\pi} \frac{\Delta q}{CV_{swing}} \tag{10.2}$$

$$\Delta\phi = \Gamma(\omega_0 \tau) \frac{\Delta V}{V_{max}} = \Gamma(\omega_0 \tau) \frac{\Delta q}{q_{max}} \tag{4.23}$$

$$\Gamma(\omega_0 \tau) = -\sin(\omega \tau) \tag{4.51}$$

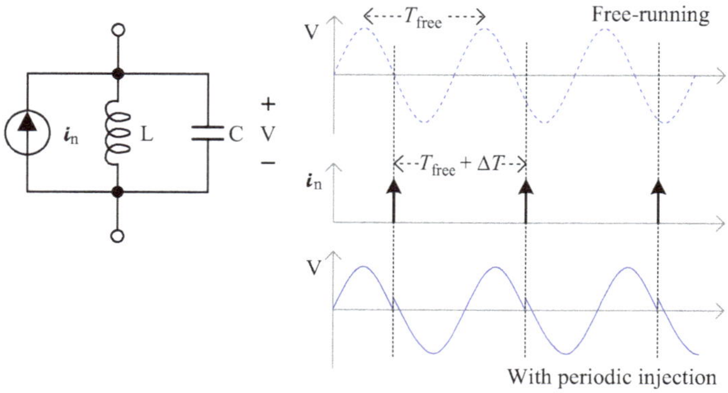

Figure 10.1 Simplified operation mechanism of the injection locking

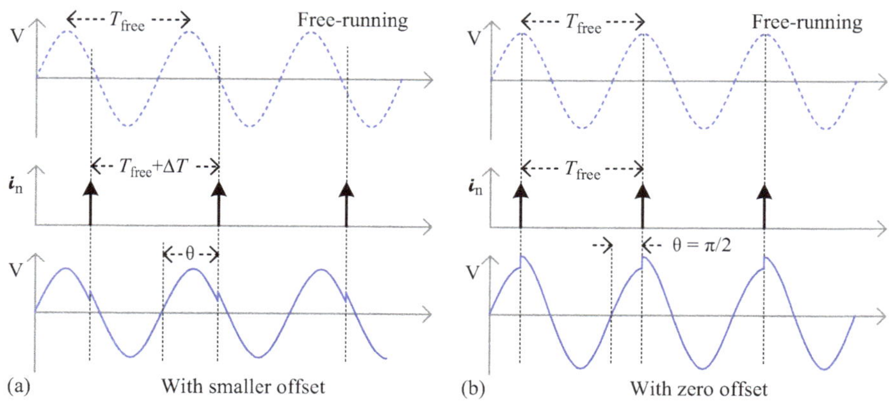

Figure 10.2 Injection locking with small frequency offset

The maximum frequency offset that can be tracked by the injection is approximately obtained as

$$\Delta f_{\max} \cong f_{\text{free}} \frac{\Delta T_{\max}}{T_{\text{free}}} = \frac{f_{\text{free}}}{2\pi} \frac{\Delta q}{q_{\max}} = \frac{f_{\text{free}}}{2\pi} \frac{\Delta q}{\text{CV}_{\text{swing}}} \tag{10.3}$$

which is referred to the lock range of the ILO. Note that Figure 10.1 shows the case that the pulse is injected to where the absolute value of the ISF is maximized. In other words, it shows the maximum frequency offset that the injection locking can exhibit. With a smaller frequency offset, assuming the same amount of charge injection, the oscillator shifts the phase where the injection happens with a smaller ISF, where the phase shift by the injection equals to the ΔT, as shown in Figure 10.2(a). For an extreme case, Figure 10.2(b) shows the waveform of the ILO when the injected frequency matches exactly to the free-running frequency. In this case, the phase shift should be nullified. Therefore, the phase of the oscillator is set to $\pi/2$ relative to the zero-crossing of the injection signal, where the ISF is zero. As observed in Figures 10.1 and 10.2, as the frequency offset increases, the phase of the oscillator becomes larger from $\pi/2$ to π.

Figure 10.3 shows the phase relation between the ILO and the injection pulse train of the opposite case, that is, the frequency offset is negative (the injection frequency is higher than the free-running frequency). Similar to the case of the slow injection, for the maximum frequency offset the injection happens when the absolute of the ISF is maximized, thus the output phase is aligned to the injection pulse. Intuitively, with a smaller frequency offset, the phase should locate somewhere between 0 and $\pi/2$. In other words, the observation of Figures 10.2 and 10.3 implies that a phase shift whose absolute value is less than $\pi/2$ is introduced when the frequency of a free-running oscillator is shifted and locked by a periodic injection. This injection-locking phenomenon was first theoretically analyzed by Robert Adler in 1946 [1].

On the other hand, if the frequency offset is larger than Δf_{max}, the phase shift is not able to compensate the cycle slip of ΔT even with the maximum ISF. Accordingly, the clock period cannot be extended to $T_{free} + \Delta T$ but periodically wanders around $T_{free} + \Delta T$ and T_{free}.

10.2 Jitter transfer of ILO

Since the injection signal affects the output phase, it is important to study the phase transfer function from the injection phase to the output phase. Assuming that the pulse train is injected, (4.23) leads to the phase transfer function:

$$\frac{\Delta\phi_{out}}{\Delta\phi_{in}} = \Gamma'(\omega_0\tau)\frac{\Delta q}{q_{max}} \tag{10.4}$$

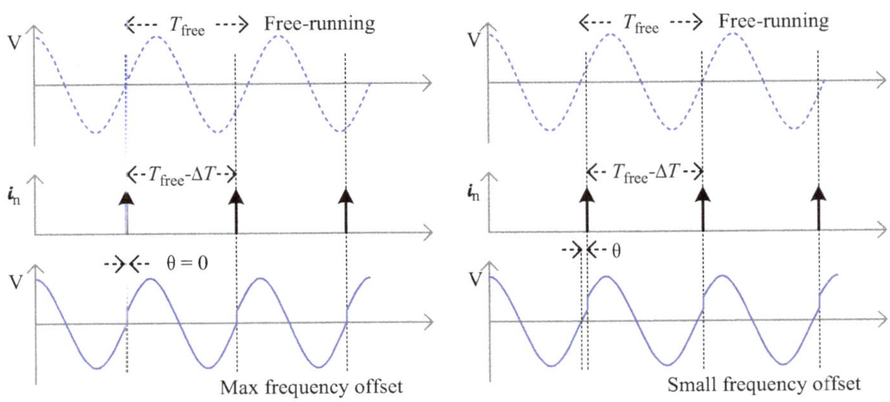

Figure 10.3 Injection locking when the injection frequency is higher than the free-running frequency

where $\Delta\phi_{in}$ and $\Delta\phi_{out}$ are the phase deviation (jitter) of the injection input and the output clock, respectively. Equation (10.4) is referred to as a realignment factor of the ILO, which is generally written as β [2,3]. Assuming the oscillator has a sinusoidal waveform, (10.4) is simplified to

$$\beta = \frac{\Delta\phi_{out}}{\Delta\phi_{in}} = \sin(\omega_0\tau)\frac{\Delta q}{q_{max}} \tag{10.5}$$

From (10.5), we can find that $0 < \beta < 1$, where $\beta = 0$ represents zero phase realignment but $\beta = 1$ means that the output phase fully tracks the input phase. Note that β is proportional to the differentiation of the ISF. Therefore, β is maximized when the injection happens and when the ISF is at the maximum (or minimum), which is relevant to the zero-frequency offset case as shown in Figure 10.2(b). Recalling (6.20), where we found that the jitter tracking bandwidth (JTB) over the reference frequency is proportional to $\frac{\Delta\phi_{out}}{\Delta\phi_{in}}$, we can obtain the JTB of an ILO as a function of β as

$$f_c = \frac{\beta}{2} f_{ref} \tag{10.6}$$

When $\beta = 1$, the theoretically maximum JTB of an ILO is obtained as ~1/2 of the reference frequency. We can find very interesting and useful nature of ILO; depending on the value of β, the ILO achieves much higher JTB than a phase-locked loop (PLL). Recall that the maximum JTB of the PLL is limited to 1/10 of the reference frequency due to the stability issue. That means, as long as the input jitter is clean, the ILO is able to suppress the oscillator phase noise a bit more.

10.3 Subharmonic ILO

To be adopted in practical applications, the ILO should generate higher output frequency from a relatively lower reference frequency, similar to PLLs. An ILO which is locked at the frequency of Nf_{ref}, where N is defined as a multiplication factor of an ILO, is referred to as subharmonic ILO, while the example given in Section 10.2 is referred to as fundamental ILO (fundamental injection). During the fundamental injection, every single edge of the oscillator clock is aligned by the injection signal. Therefore, the phase of each edge is set by the same mechanism, so the spacing between adjacent edges equals each other when we neglect other nonidealities (i.e., input jitter and oscillator phase noise). However, the subharmonic injection corrects once every N edges. As a result, the phase of the edge affected by the injection is set by a different mechanism (the combination of the

---

$$\Delta\phi = \Gamma(\omega_0\tau)\frac{\Delta V}{V_{max}} = \Gamma(\omega_0\tau)\frac{\Delta q}{q_{max}} \tag{4.23}$$

$$\frac{\Delta\phi_{out}}{\phi_{err}} \cong \frac{2\omega_c}{\omega_{ref}} = \frac{2f_c}{f_{ref}} \tag{6.20}$$

free-running and the injection, same as the fundamental injection) to the other edges. Because there is no injection during the rest of the cycles, the other edges are solely set by the oscillator's free-running mechanism. As a result, the time interval of those cycles equals to T_{free}, whereas that influenced by the injection deviates to $T_{\text{free}} + \Delta T$. Figure 10.4 illustrates an example of the subharmonic injection, where the free-running frequency is at the vicinity of 3× of the injection frequency. Because the injection frequency is 1/3 of the locked frequency, the frequency of the output clock is periodically fluctuated. We have already studied a similar phenomenon of the reference spur by the PLL nonidealities. That is, we can infer that the subharmonic injection also leads to the reference spur, if there is an offset between the subharmonic of the free-running frequency and the injection frequency.

Figure 10.5 shows the phase deviation of the ILO clock under the subharmonic injection. The phase is periodically modulated, while the peak-to-peak amplitude of the modulation is calculated as

$$2\Delta\phi = 2\pi\frac{\Delta T}{T_{\text{free}} + \Delta T/N} = 2\pi\frac{\alpha}{1 + \alpha/N} \approx 2\pi\alpha \tag{10.7}$$

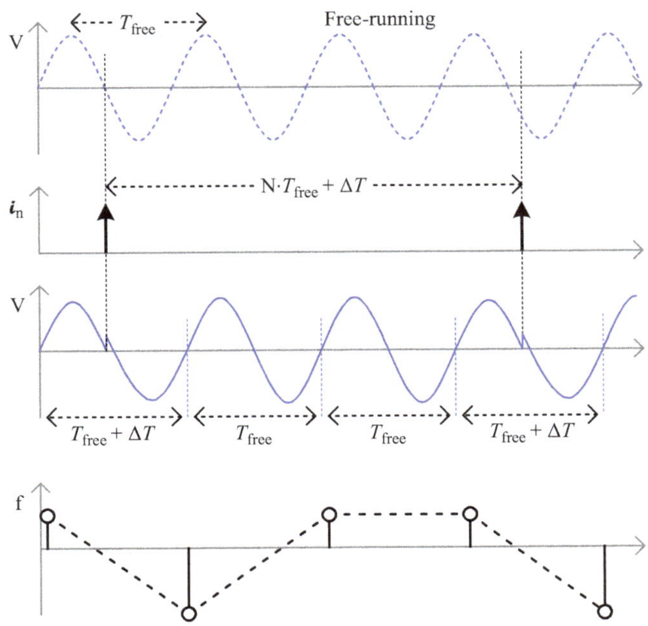

Figure 10.4 Clock waveform and frequency under the subharmonic injection

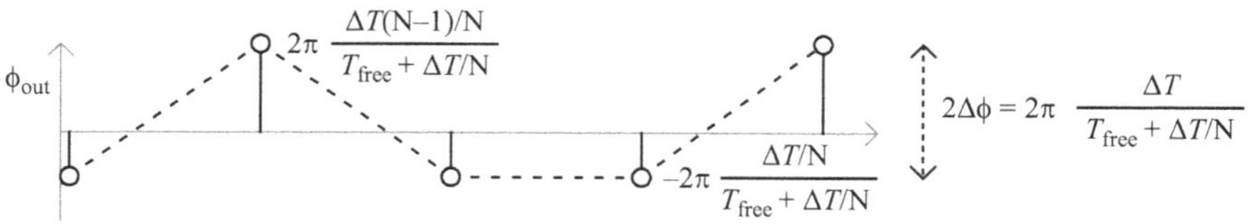

Figure 10.5 Clock phase deviation due to the subharmonic injection

where α is the normalized frequency offset ($\Delta f/f_{free}$ or $\Delta T/T_{free}$). Recalling (2.17) and Figure 2.4, the signal power of a small phase modulation (amplitude $= \Delta\phi$) that is normalized to the power of the carrier signal is expressed as

$$10\log\left(\frac{P_{spur}}{P_{carrier}}\right) = 10\log\left(\frac{\Delta\phi^2}{2}\right) = 20\log\left(\frac{\Delta\phi}{2}\right) \approx 20\log\left(\frac{\pi\alpha}{2}\right) \tag{10.8}$$

Substituting (10.3) and (10.4), the maximum value of α is also expressed as a function of β:

$$\alpha_{max} = \frac{\Delta f_{max}}{f_{free}} = \frac{1}{2\pi}\frac{\Delta q}{q_{max}} = \frac{1}{2\pi}\frac{\beta}{\Gamma'(\omega_0\tau)} \tag{10.9}$$

For example, with the injection strength of 0.1, the worst-case spur due to the injection is calculated as –32 dBc, which is very poor. For example, using (6.69), this amount of spur is relevant to the reference spur from the I_{leak} of $0.057I_{CP}$. Note that the spur should become zero if the free-running frequency matches well with the injection frequency. From this observation, we can find that a proper tuning of the free-running frequency is very important to suppress the reference spur.

10.4 ILO circuit implementation

This chapter will introduce a few popular circuit implementations of ILO. Figure 10.6 shows the most famous injection scheme, where a switch shorts the differential node of the oscillator when the injection pulse is triggered [4–11]. Depending on the strength of the switch, the differential amplitude of the clock gets smaller. Therefore, if the injection happens ahead of the zero-crossing of the free-

---

$$\begin{aligned} S_y(f) = {} & \frac{1}{4}A_0^2\delta(f-f_0) + \frac{1}{4}A_0^2\delta(f+f_0) \\ & + \frac{1}{4}A_0^2\left(\frac{1}{4}A_1^2\delta(f-f_0-f_1) + \frac{1}{4}A_1^2\delta(f-f_0+f_1)\right. \\ & \left. + \frac{1}{4}A_1^2\delta(f+f_0-f_1) + \frac{1}{4}A_1^2\delta(f+f_0+f_1)\right) \end{aligned} \tag{2.17}$$

$$\text{Spur} = 20\log\left(\frac{2}{\pi}T(f_{ref})\cdot\sin\left(\pi\frac{I_{leak}}{I_{CP,0}}\right)\right) \tag{6.69}$$

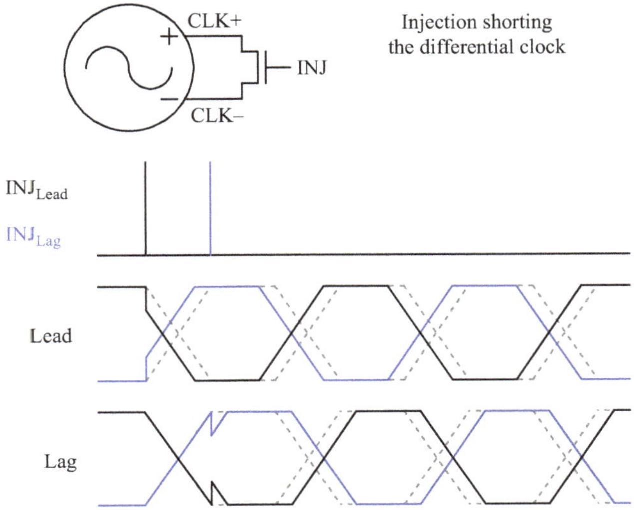

Figure 10.6 ILO implementation with a shorting switch

running clock, the transition is accelerated so the injection brings the zero-crossing forward. On the other hand, the injection after the zero-crossing interrupts the transition, so it delays the next zero-crossing. As a result, under injection-locked, the injection takes place before the zero-crossing to accelerate the oscillator when the injection frequency is faster than the free-running frequency and vice versa.

One of the most important design issues in such switch-based ILO is making the pulse narrow. Note that we assumed impulse injection in the study above. Reference [8] showed that the finite pulse-width introduces asymmetric phase response as shown in Figure 10.7. That is, as the pulse gets wider, it becomes harder to obtain the negative phase shift. It degrades the lock range and the jitter transfer. In extreme case, the ILO is not able to achieve the negative phase shift, which means that the ILO cannot be locked even when the injection frequency equals to the free-running frequency.

For the case of subharmonic injection which is the main interest of this study, the realignment factor (β) also depends on the pulse-width. While (10.5) shows that the β is primarily proportional to Δq, [7] verifies that β is also affected by the pulse-width. Assuming a pulse train $P(t)$ whose pulse-width and frequency are D and $1/f_{ref}$, respectively, we can write $P(t)$ using Fourier series as

$$P(t) = \frac{\Delta q}{D}\left\{ a_0 + \sum_{n=1}^{\infty} a_n \cos\left(2\pi f_{ref} \cdot n \cdot t\right) \right\} \tag{10.10}$$

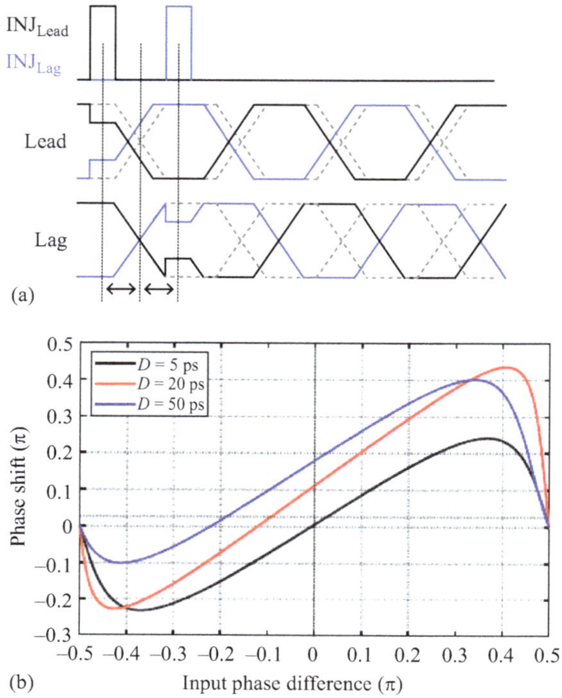

Figure 10.7 (a) ILO with wide injection pulse and (b) asymmetric phase response due to the wide pulse-width

where

$$a_0 = f_{\text{ref}} \cdot D \tag{10.11}$$

$$a_n = \frac{2}{n\pi} \sin\left(\pi f_{\text{ref}} \cdot n \cdot D\right) \tag{10.12}$$

Here, assume that a sinusoidal injection leads to a sinusoidal phase modulation. Recalling Section 2.2.2, a sinusoidal phase modulation introduces the spectral tones at $\pm f_{\text{mod}}$ apart from the carrier frequency (f_{VCO} for the ILO case). From this property, we can observe that if the modulation frequency is $2f_{\text{VCO}}$, it adds up the signal power at f_{VCO}. From the definition of the phase noise given in (2.22), the phase noise is improved as the power at the carrier frequency increases. With the multiplication factor of N, the amplitude of sinusoidal component at $2f_{\text{VCO}}$ is rewritten as

$$\frac{\Delta q}{D} \cdot a_n = \frac{\Delta q}{D} \cdot \frac{2}{2N\pi} \cdot \sin\left(\pi f_{\text{ref}} \cdot 2N \cdot D\right) = \frac{\Delta q}{D} \cdot \frac{1}{N\pi} \cdot \sin\left(2\pi f_{\text{VCO}} \cdot D\right) \tag{10.13}$$

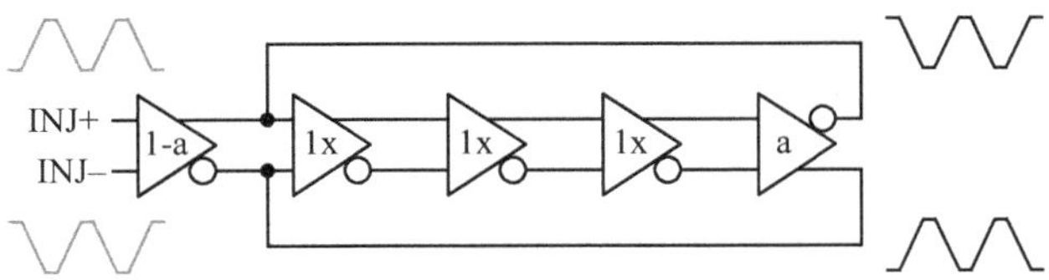

Figure 10.8 ILO implementation with direct injection

where we find that $D = T_{VCO}/4$ maximizes the harmonic power at $2f_{VCO}$ (or $2Nf_{ref}$) assuming a fixed Δq, which has been proven in [7] with the fabrication result. At the observation from Figure 10.7 and (10.13) [7, 8], it can be concluded that the injection pulse-width should be narrow (at least as narrow as $T_{VCO}/4$). Note that we need to generate such narrow pulse from the reference clock in the chip. Therefore, the oscillator clock is no longer the fastest signal in the chip. Instead, the injection signal becomes the fastest, so the speed of an ILO is constrained by the injection-pulse generator circuit. The direct injection shown in Figure 10.8 can be an alternative since the input clock is directly injected to the ILO hence it does not require a pulse generator. However, the direct injection is not appropriate to support the clock multiplication with the subharmonic injection, so its application is a bit limited. AC coupling can be used to isolate the common-mode of the injection buffer.

From now on, we will study how the narrow injection pulse can be generated. The most straightforward implementation might be XOR-gate-based pulse generator as shown in Figure 10.9. Note that the injection pulse is generated from both rising and falling edges of the reference clock. Since the effective injection frequency is doubled, the lock range and the JTB are ideally doubled. Whereas the main advantage is simple implementation and doubled injection frequency, there is a critical disadvantage that it is vulnerable to the duty-cycle distortion of the reference clock. Because rising and falling edges are used, the amount of the duty-cycle distortion directly propagates to the pulse train, as shown in Figure 10.9(b). If there is the duty-cycle distortion, the even pulses and the odd pulses are injected at a different cycle each other. As a result, even though there is no frequency mismatch between the free-running frequency and the injection frequency, the ILO ends up where the injection induces phase shifts. For example, as shown in Figure 10.10, the even and odd injections introduce the negative and positive phase shifts, respectively, to nullify the sum of the shifts. Assuming that the duty-cycle is distorted by δ (in time), the ILO output is shifted by $\beta{\cdot}\delta$ as definition of β, which leads to the peak-to-peak deterministic jitter of $2\beta{\cdot}\delta$. Noting that the ILO frequency is N- times higher than the reference frequency in general, the duty-cycle distortion is translated to the deterministic jitter with the amplification factor of N. For

---

$$\mathcal{L}(\Delta f) = 10\log\left(\frac{P_{\text{sideband}}(f_0+\Delta f,\ 1\text{ Hz})}{P_{\text{carrier}}}\right) \tag{2.22}$$

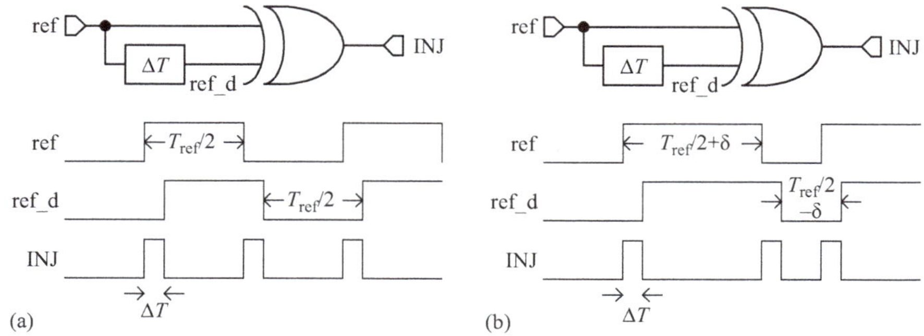

Figure 10.9 XOR-gate-based pulse generator and timing diagram: (a) without duty-cycle distortion and (b) with duty-cycle distortion

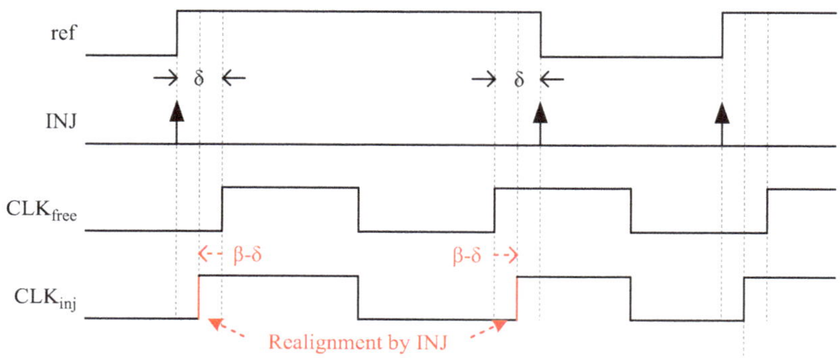

Figure 10.10 Deterministic jitter induced by the duty-cycle distortion of the reference clock

example, even if the duty-cycle distortion looks negligible at the reference clock (i.e., 0.001UI), it will not be negligible with a high multiplication factor N (i.e., 64). Assuming a moderate β of 0.25, we get the deterministic jitter of 0.032 UI_{pp}, which is substantial. Therefore, high-precision duty-cycle correction is highly required for the reference clock [11]. From (2.17), we can also calculate the spur level from the deterministic jitter as

$$\text{Spur} = 20\log\left(\frac{A_1}{2}\right) \approx 20\log\left(\frac{2\pi DJ}{2T_{VCO}}\right) = 20\log\left(\frac{\pi\cdot\beta\cdot\delta}{T_{VCO}}\right) \tag{10.14}$$

where A_1 and DJ are the amplitude of sinusoidal modulation in radian and the amplitude of the deterministic jitter (half of the peak-to-peak), respectively. Note that we assume the deterministic jitter in radian equals A_1 for simplicity.

In order to avoid the aforementioned duty-cycle issue, majority of the recent works use a frequency divider preceding to the XOR-based pulse generator or

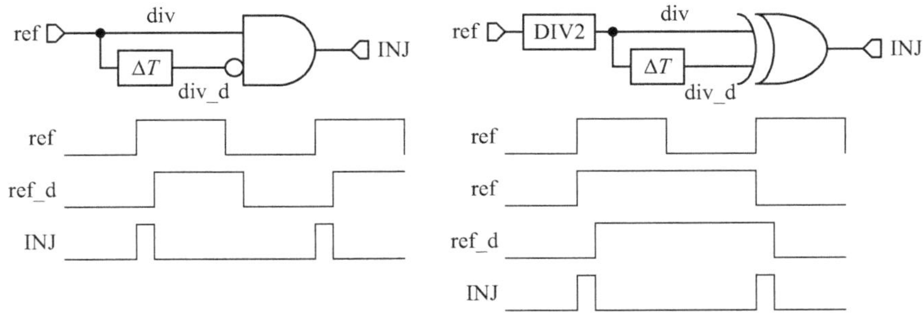

Figure 10.11 AND-gate-based pulse generator and XOR-based pulse generator preceded by a frequency divider

AND-gate-based pulse generator [8–10,12] (Figure 10.11). While sacrificing higher JTB and lock range, it eliminates the deterministic jitter by the duty-cycle error.

On the other hand, the pulse-generator approach has a bandwidth issue as we have already briefly discussed. Since the XOR gate produces narrow pulse whose pulse-width is even much narrower than that of the ILO clock, this pulse is fed to the shorting switch whose load is not negligible and the required bandwidth of the XOR gate is too high, especially for high-speed clock applications. A complimentary switch (CS) injection has been proposed in [11] as an alternative. Instead of producing a short pulse, the CS injection uses the exact moment when based on a switched-capacitor operation. Figure 10.12 shows the schematic diagram of the CS scheme which is composed of four switches and a capacitor. We can find that the reference clock rather than the injection pulse is directly fed to the shorting switches. At a steady-state, that is, when the reference clock stays at a certain level (high or low), only two switches turn on. For example, while the ref+ is high, the top-left and bottom-right switches turn on, so the left plate (CS+) and the right plate (CS−) of the capacitor track the voltages of the CLK+ node (V_{CLK+}) and the CLK− node (V_{CLK-}), respectively. At the moment of the reference clock high-to-low transition, the polarity of the capacitor should be flipped. As a result, the capacitor discharges $2 \cdot C_{CS} \cdot (V_{CLK}+ - V_{CLK-})$ through the current path between CLK+ and CLK− (I_{INJ}) at the moment of the transition. The duration of the pulse is a function of the turn-on resistance of the switch. Because this scheme produces the injection current pulse directly from the reference clock whereas the pulse generator approach needs an intermediate voltage pulse, it eliminates the

---

$$\begin{aligned} S_y(f) &= \frac{1}{4}A_0{}^2\delta(f-f_0)+\frac{1}{4}A_0{}^2\delta(f+f_0) \\ &+\frac{1}{4}A_0{}^2\left(\frac{1}{4}A_1{}^2\delta(f-f_0-f_1)+\frac{1}{4}A_1{}^2\delta(f-f_0+f_1)\right. \\ &\left.+\frac{1}{4}A_1{}^2\delta(f+f_0-f_1)+\frac{1}{4}A_1{}^2\delta(f+f_0+f_1)\right) \end{aligned} \tag{2.17}$$

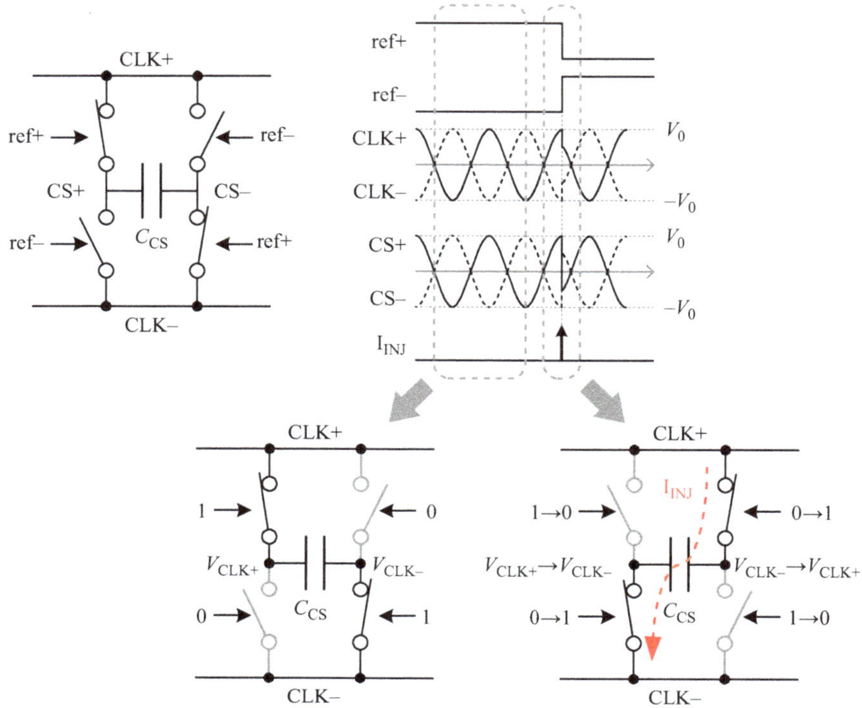

Figure 10.12 Complimentary switch injection

bandwidth issue. Since the amount of injected charges is proportional to the C_{CS}, the injection strength increases as the C_{CS} becomes larger. However, the C_{CS} should be chosen carefully because it decreases the free-running frequency of the ILO. Note that the C_{CS} is coupled to the differential nodes, hence the effective load capacitance added by the C_{CS} is amplified by the Miller effect.

10.5 Injection-locked PLL

In Sections 10.1 and 10.3, we analyzed the lock range and the spur due to the frequency offset of the ILO. These issues make it hard to use a standalone ILO for practical clocking circuits because there are many nonideal conditions such as process, voltage, and temperature (PVT) variations which never allow an oscillator to work at a fixed frequency. Therefore, an ILO is generally combined with some other circuits which push the ILO free-running frequency into the lock range and correct the residual frequency offset. Therefore, ILOs are frequently combined with PLLs, which is generally called injection-locked PLL (IL-PLL). Moreover, a standalone ILO exhibits a worse phase noise at the mid-frequency range compared to a PLL, although the ILO provides a wider JTB (or cut-off bandwidth).

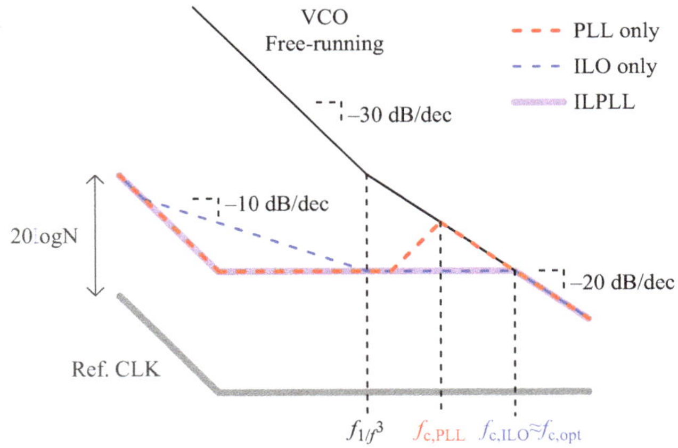

Figure 10.13 Phase noise comparison between PLL, ILO, and IL-PLL

Figure 10.13 illustrates the phase noise comparison of the PLL and the ILO, where f_{1/f^3}, $f_{c,PLL}$, $f_{c,ILO}$, and $f_{c,opt}$ represent the $1/f^3$ noise cut-off frequency, PLL's cut-off frequency (loop bandwidth), ILO's cut-off frequency, and the optimum cut-off frequency that minimizes the phase noise, respectively. For example, let us assume that we have a relatively clean reference clock but poor VCO phase noise. We also make an assumption that the PLL is not able to meet the optimum cut-off bandwidth because it is constrained by the 1/10 of the reference frequency, to justify the use of ILO. As described in Chapter 6, this introduces excessive phase noise at the vicinity the frequency range of $f_{c,PLL}$ to $f_{c,opt}$ because the VCO phase noise is not filtered sufficiently in that region. On the other hand, it is assumed that the ILO can meet the optimum cut-off frequency thanks to the wide bandwidth. However, since the ILO is a first-order low-pass filter in the phase domain, it offers filtering capacity of by only 20 dB/dec. As a result, the ILO is not enough to suppress the VCO phase noise in –30 dB/dec region, which is fully suppressed in the case of the PLL. Considering that both PLL and ILO are dominated by the $1/f$ noise of the reference clock at very low-frequency noise region, we can see that the ILO is better than the PLL in the high-frequency region but is worse in the mid-frequency region. Here we can find another main motivation of the IL-PLL; we can take the mid-frequency performance of the PLL as well as the high-frequency performance of the ILO by combining the ILO with the PLL. From now on, we will study some practical issues raised when combining them and state-of-the-art examples of IL-PLLs.

Figure 10.14(a) shows a representative block diagram of an IL-PLL. Because the PLL brings the output frequency to the desired frequency (Nf_{ref}), of course the ILO meets the lock range requirement. Since the PLL also tracks the PVT variations, the output frequency is always fixed so it seems there is no deterministic jitter caused by the injection with a residual frequency offset. However, in practice, finite delays in both the PLL feedback path (D_{FB}) and the injection path (D_{inj}) introduce a

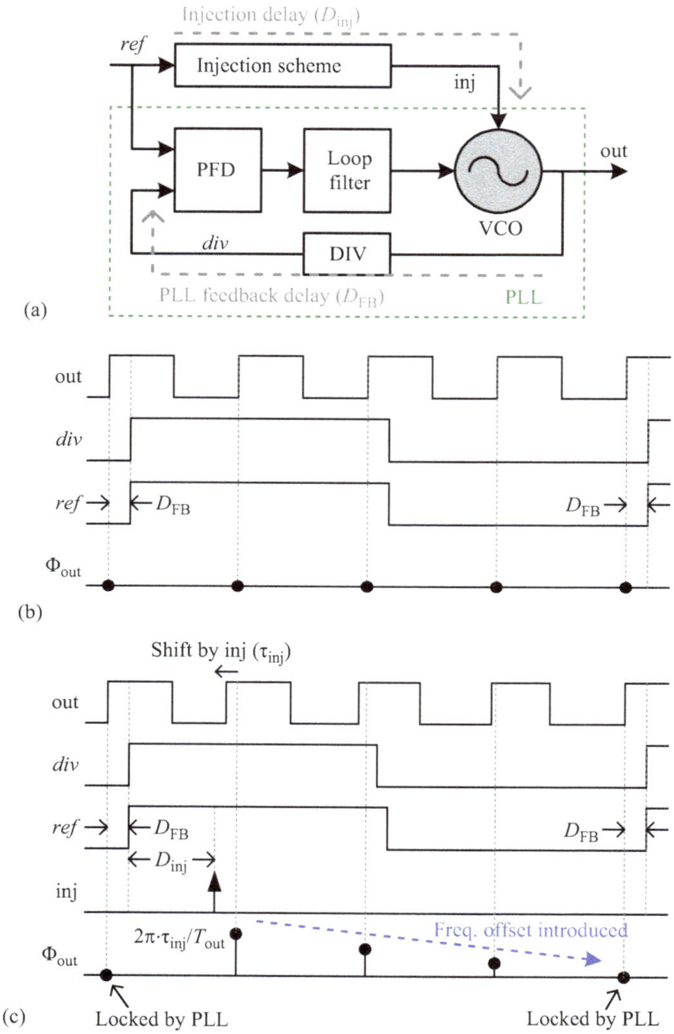

Figure 10.14 (a) Block diagram of IL-PLL, (b) timing diagram without injection (PLL only), and (c) timing diagram with injection

deterministic jitter. Figure 10.14(b) shows the timing diagram of the IL-PLL when the injection is disabled (PLL-only configuration). Because the PLL locks the phase and frequency of the divided clock (*div* in Figure 10.14) with those of the reference clock (*ref* in Figure 10.14), the output clock has an offset of $-D_{FB}$ with respect to the *ref* when the PLL is locked. Neglecting all noise sources, the output clock has zero period jitter even though the PLL corrects the output phase at every N cycle. However, while being enabled, the injection introduces another shift (τ_{inj})

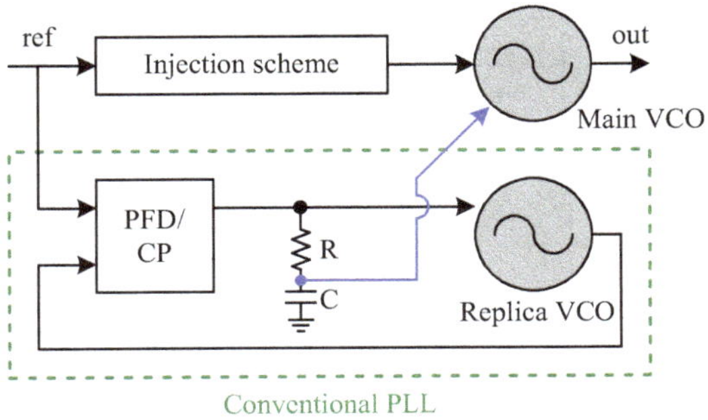

Figure 10.15 IL-PLL with a replica oscillator

with respect to the time difference between the injection pulse and the closest rising edge of the output clock. If the *k*th rising edge is the closest one, the time difference can be written as

$$\Delta t = k \cdot T_{\text{out}} - D_{\text{inj}} - D_{\text{FB}} \tag{10.15}$$

Then the shift by the injection becomes:

$$\tau_{\text{inj}} = \beta\Delta t = \beta \cdot \left(k \cdot T_{\text{out}} - D_{\text{inj}} - D_{\text{FB}}\right) \tag{10.16}$$

However, there is a PLL which is assumed to force the *ref* and *div* to be aligned at every rising edge. If the free-running frequency is perfectly matched to the injection frequency, the τ_{inj} propagates forever so the PLL can never be locked. That means, the IL-PLL should be locked only when there is a frequency offset. Again, injection locking with a frequency offset results in the reference spur. The spur level is approximated from (10.14) as

$$\text{Spur} \approx 20 \log\left(\frac{\pi \cdot \beta \cdot \left(k \cdot T_{\text{out}} - D_{\text{inj}} - D_{\text{FB}}\right)}{2T_{\text{out}}}\right). \tag{10.17}$$

From (10.17), we can find that the spur is eliminated when

$$D_{\text{inj}} + D_{\text{FB}} = k \cdot T_{\text{out}} \tag{10.18}$$

Because the frequency offset in IL-PLL is caused by the conflict between two phase alignment forces (PLL and ILO), [5] has proposed to use a replica oscillator in order to isolate the ILO from the PLL loop. In Figure 10.15, the PLL incorporates a replica oscillator instead of having the main ILO in the loop. The ILO takes its control voltage from the loop filter capacitor which stores only the frequency information, and therefore it obtains the frequency lock from the PLL but not the phase lock. As a result, the phase of the main VCO is solely controlled by the injection independent to the PLL. The main disadvantage of this architecture is doubled power and area consumption, because two oscillators are in use at the same

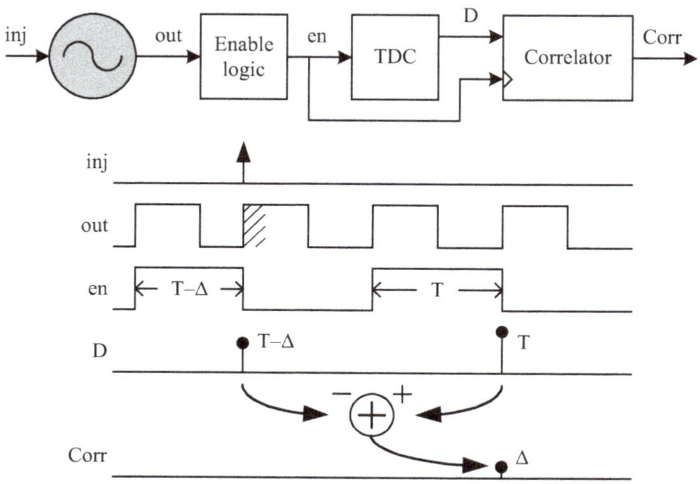

Figure 10.16 TDC-based injection timing correction circuit

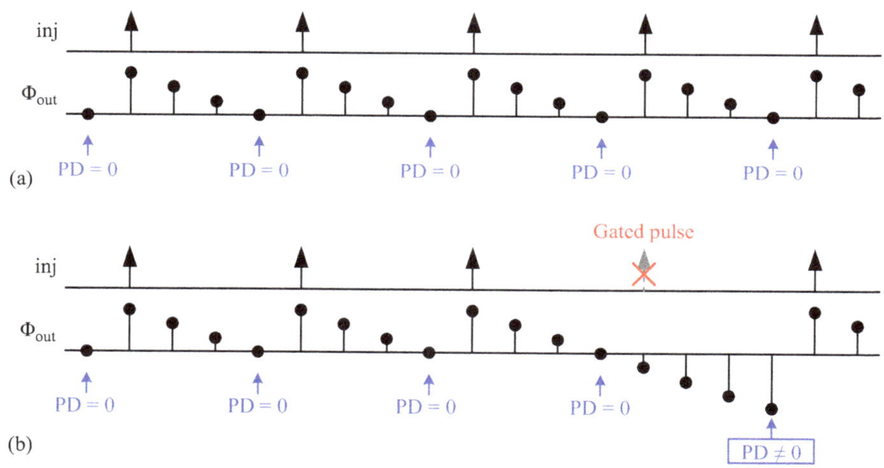

Figure 10.17 Conceptual timing diagram of gated pulse technique: (a) timing diagram of typical IL-PLL and (b) timing diagram of gated-pulse IL-PLL

time. Moreover, since this architecture works well only when those two VCOs are matched very well, the VCOs need to be even larger in order to reduce the mismatch.

An alternative way is inserting a variable delay line in the injection path (or in the feedback path) to meet (10.18). The variable delay is tuned to nullify the frequency offset. In this approach, the main design challenge becomes how to collect the frequency offset information. Reference [4] pays attention to the pulse-width distortion by the injection. From the out waveform in Figure 10.14(c), we can observe that the duration of the first negative pulse is reduced due to the injection. Based on this observation, [4] proposes to compare the output pulse width using a high-resolution TDC (Figure 10.16). The TDC measures the free-running period and the injection-affected period. A digital correlator is used to subtract the two measured results. By accumulating the subtracted value (Δ), the misalignment information is collected and is used to adjust the variable delay, which is similar to duty-cycle correction circuits. Moreover, note that this approach is free from any mismatch issue because the period is measured by the same device and the same path. However, the accuracy of this correction is limited by the finite TDC resolution, and moreover the high-resolution TDC consumes huge power and large area. There are some other state-of-the-art works that adopt this kind of pulse-width-based approach [9,10,13], in which they tried to measure the pulse-width without using the power-hungry TDC.

A gated pulse technique proposed in [8] collects the frequency offset information without introducing power-hungry circuits. As studied in Figure 10.14(c), an IL-PLL locks with a frequency offset which compensates the phase shift introduced by the injection. Therefore, the output of the PLL phase detector is always zero, which means that the PLL does not recognize the frequency offset, as shown in Figure 10.17(a). The idea of the gated pulse is intentional omission of one of the injection pulses with a pulse gating circuit. Once one of the pulses is gated, the phase drift by the frequency offset has no chance to be corrected by the injection at that cycle. As a result, there will be nonzero phase difference at the moment of the right next phase detection of the omission, as shown in Figure 10.17(b). From the result from the PLL phase detector, we can extract the frequency offset. Although this technique successfully tracks the frequency offset, the specific interval for the pulse gating leads to a fractional spur. In addition, the max achievable JTB is degraded depending on the gating rate.

References

[1] Adler R. A study of locking phenomena in oscillators. Proceedings of the IRE. 1946;34(6):351–357.

[2] Elkholy A, Talegaonkar M, Anand T, Hanumolu PK. Design and analysis of low-power high-frequency robust sub-harmonic injection-locked clock multipliers. *IEEE Journal of Solid-State Circuits*. 2015;50(12):3160–3174.

[3] Gierkink SL. Low-spur, low-phase-noise clock multiplier based on a combination of PLL and recirculating DLL with dual-pulse ring oscillator and self-correcting charge pump. *IEEE Journal of Solid-State Circuits*. 2008; 43(12):2967–2976.

[4] Helal BM, Hsu CM, Johnson K, Perrott MH. A low jitter programmable clock multiplier based on a pulse injection-locked oscillator with a highly-

digital tuning loop. *IEEE Journal of Solid-State Circuits*. 2009;44(5): 1391–1400.

[5] Musa A, Deng W, Siriburanon T, Miyahara M, Okada K, Matsuzawa A. A compact, low-power and low-jitter dual-loop injection locked PLL using all-digital PVT calibration. *IEEE Journal of Solid-State Circuits*. 2014; 49(1):50–60.

[6] Chien JC, Upadhyaya P, Jung H, *et al.* 2.8 A pulse-position-modulation phase-noise-reduction technique for a 2-to-16 GHz injection-locked ring oscillator in 20nm CMOS. In *2014 IEEE International Solid-State Circuits Conference Digest of Technical Papers (ISSCC)*. San Francisco, CA: IEEE; 2014 (pp. 52–53).

[7] Wei CL, Kuan TK, Liu SI. A subharmonically injection-locked PLL with calibrated injection pulsewidth. *IEEE Transactions on Circuits and Systems II: Express Briefs*. 2015;62(6):548–552.

[8] Elkholy A, Talegaonkar M, Anand T, Hanumolu PK. Design and analysis of low-power high-frequency robust sub-harmonic injection-locked clock multipliers. *IEEE Journal of Solid-State Circuits*. 2015;50(12):3160–3174.

[9] Choi S, Yoo S, Lim Y, Choi J. A PVT-robust and low-jitter ring-VCO-based injection-locked clock multiplier with a continuous frequency-tracking loop using a replica-delay cell and a dual-edge phase detector. *IEEE Journal of Solid-State Circuits*. 2016;51(8):1878–1889.

[10] Kim S, Ko HG, Cho SY, *et al.* 29.7 A 2.5 GHz injection-locked ADPLL with 197fs rms integrated jitter and −65 dBc reference spur using time-division dual calibration. In *2017 IEEE International Solid-State Circuits Conference (ISSCC)*. San Francisco, CA: IEEE; 2017 (pp. 494–495).

[11] Cho SY, Kim S, Choo MS, *et al.* A 2.5–5.6 GHz subharmonically injection-locked all-digital PLL with dual-edge complementary switched injection. *IEEE Transactions on Circuits and Systems I: Regular Papers*. 2018; 65(9):2691–2702.

[12] Lee J, Wang H. Study of subharmonically injection-locked PLLs. *IEEE Journal of Solid-State Circuits*. 2009;44(5):1539–1553.

[13] Raj M, Saeedi S, Emami A. A wideband injection locked quadrature clock generation and distribution technique for an energy-proportional 16–32 Gb/s optical receiver in 28 nm FDSOI CMOS. *IEEE Journal of Solid-State Circuits*. 2016;51(10):2446–2462.

Chapter 11
Phase noise suppression techniques 4: clock multiplying DLL

11.1 DLL with an edge-combining logic

Frequency multiplication with a DLL is not as easy as that with a PLL; however, there have been a lot of efforts to adopt DLLs for frequency multiplication, in order to take advantage of its better jitter performance owing to less jitter accumulation. A primitive way is to utilize multiphase clock that can be generated from a type-I DLL [1]. The basic concept is that equally spaced phases at the reference frequency are processed through an edge-combining logic. The simplest example is a frequency doubler shown in Figure 11.1. Basically, we have enough edge information from the original phase (clk0) and the DLL-generated phase (clk90), we can produce a doubled frequency. However, any mismatch in the delay elements (i.e., D0 $\neq$ D1) or the edge-combining logic (i.e., two inputs of the XOR) directly translates into duty-cycle error and deterministic jitter at the output clock. Moreover, programmable multiplication ratio is difficult to achieve.

11.2 Multiplying DLL

A multiplying DLL (MDLL), which is also referred as to a recirculating DLL, is another option to utilize a DLL for frequency multiplication. In fact, it is not a literal DLL, because the delay line in an MDLL forms a negative feedback, and therefore the clock pulse circulates through the delay line during most of the time of operation. As a result, there is jitter accumulation. However, the loop is the first order so that it only controls the delay of the line, which is similar to a DLL. In fact, it stands on somewhere between PLL and DLL. Maybe a better way to understand MDLL is, considering it as an ILO with realignment factor of 1. Figure 11.2 shows a basic MDLL implementation and operation. It looks similar to a DLL (or a PLL), but a MUX and a selection logic are added [2,3]. When the selection signal (SEL) is "0," the output of the VCDL is fed back to its input, so it becomes a ring VCO (VCO mode). During this cycle, the jitter is accumulated through the ring and the MDLL works in a similar way of a PLL. But note that the loop filter does not have a resistor, which takes charge of the phase control (and stability!). Therefore, it is more like a frequency-locked loop. The selection signal stays at "0" during most of the time, but the selection logic toggles it to "1" at the vicinity of the rising edge of

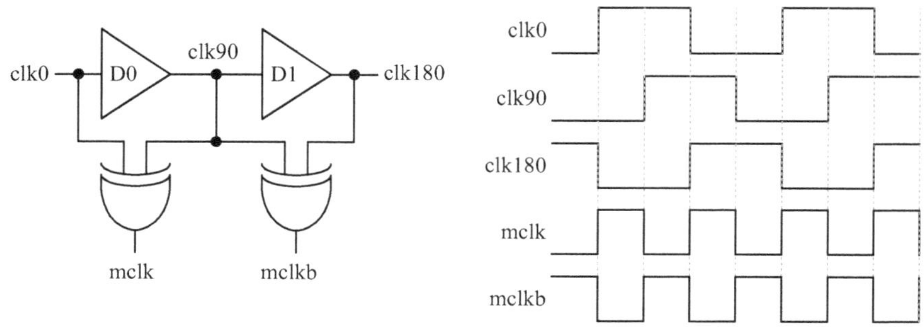

Figure 11.1 Frequency doubler based on DLL

the reference clock (*REF*). Then the VCO feedback is broken and the reference clock edge is directly inserted to the VCDL (VCDL mode). After that, the *SEL* returns to "0" so the MUX forms the VCO again. Therefore, the VCO starts oscillating with a clean reference edge. We can find that it is similar to a PLL and a subharmonic ILO, where contaminated edges by the jitter accumulation are corrected at every rising edge of the reference clock. However, recalling that (6.20) and (10.5), the realignment factor of a PLL or an ILO is less than unity, whereas that of MDLL is unity since the reference edge is directly inserted to the VCO. Substituting $\beta = 1$ to (10.6) leads to the JTB of MDLL to $f_c = \frac{1}{2}f_{\text{ref}}$, which is ~5× higher than the PLL limit. As a result, a MDLL offers much better capability to suppress oscillator phase noise:

Figure 11.2(b) gives a more detailed explanation of how the MDLL achieves lock. The first rising edge of the *REF* is injected to the VCDL through MUX (D_{MUX} is the delay of the MUX). During the first reference period, the control voltage (V_{ctrl}) is initially set to a bit high so the oscillator runs faster than the reference clock in the given example. As a result, the divided clock (*DIV*) leads the second rising edge of the *REF*. And then, the phase detector and the charge-pump decrease the V_{ctrl} proportional to the phase difference. This correction continuous to until the *REF* and the *DIV* are phase-locked.

The MDLL operation shown in Figure 11.2 is assumed to be ideal; however, path mismatches in the MUX and the phase detector cause nonideality (Figure 11.3) [4–6]. These path mismatches also happen in a PLL, but the outcome

---

$$\frac{\Delta\phi_{\text{out}}}{\phi_{\text{err}}} \cong \frac{2\omega_c}{\omega_{\text{ref}}} = \frac{2f_c}{f_{\text{ref}}} \tag{6.20}$$

$$\beta = \frac{\Delta\phi_{\text{out}}}{\Delta\phi_{\text{in}}} = \sin(\omega_0\tau)\frac{\Delta q}{q_{\max}} \tag{10.5}$$

$$f_c = \frac{\beta}{2}f_{\text{ref}} \tag{10.6}$$

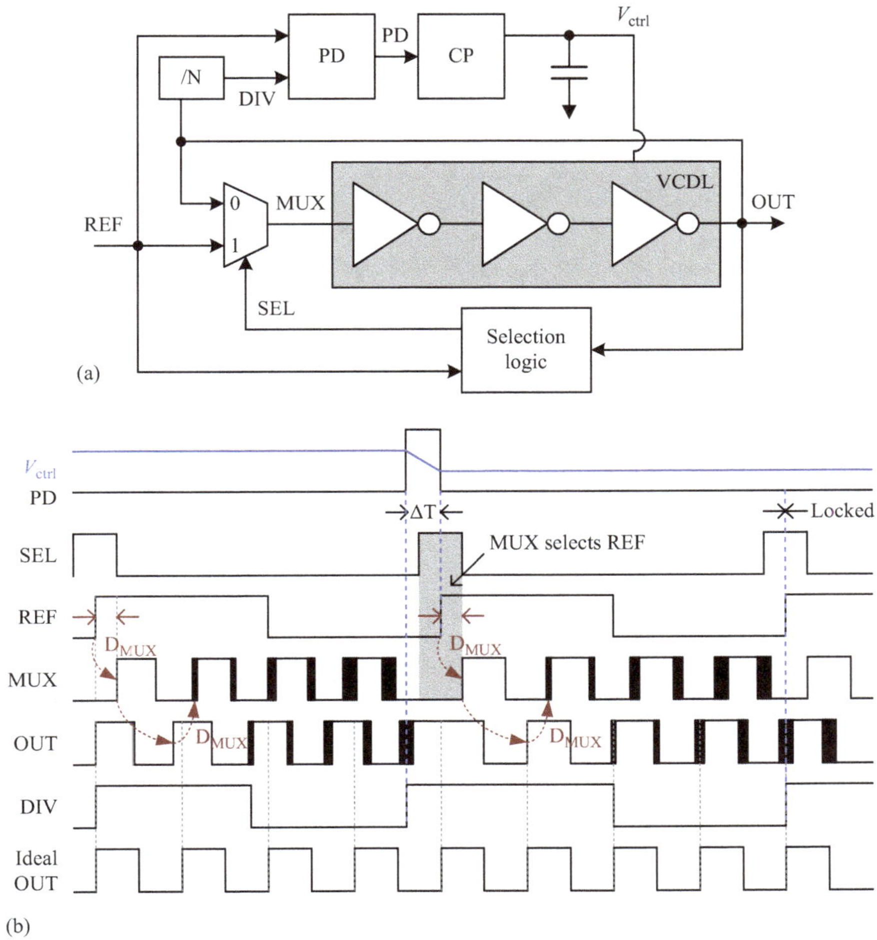

Figure 11.2 (a) Block diagram and (b) waveforms of basic MDLL

is a bit different [6]. Figure 11.4(a) shows an example of waveforms and output jitter of an MDLL when the phase detector has an offset between two input paths. At the steady-state, we can split the reference period (T_{REF}) into:

$$T_{REF} = \left(D_{MUX,1} + 2D_{VCDL} + D_{MUX,0}\right) + (2N - 1)\left(D_{VCDL} + D_{MUX,0}\right) + \Delta t_{PD} \tag{11.1}$$

where $D_{MUX,0}$, $D_{MUX,1}$, D_{VCDL}, and Δt_{PD} represent the MUX delay of each input, the VCDL delay, and the phase detector offset. By substituting $D_{MUX,1}$ with $D_{MUX,0} + \Delta D_{MUX}$, (11.1) is simplified to

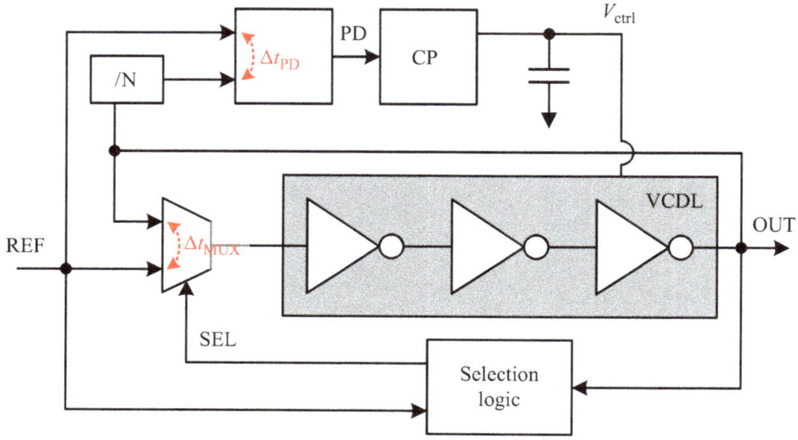

Figure 11.3 Offset introduced by path mismatches in MDLL

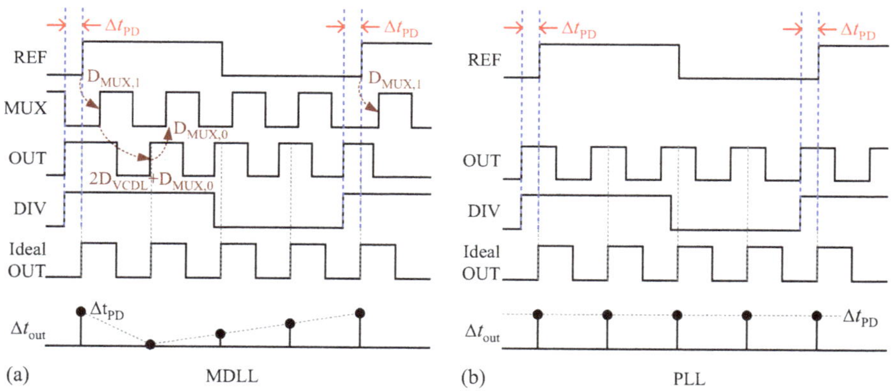

Figure 11.4 Output jitter with the phase detector mismatch of (a) MDLL and (b) PLL

$$T_{REF} = 2N\left(D_{VCDL} + D_{MUX,0}\right) + \Delta D_{MUX} + \Delta t_{PD} \tag{11.2}$$

We can obtain the period of oscillator as

$$T_{VCO} = 2\left(D_{VCDL} + D_{MUX,0}\right) = \frac{T_{REF} - \Delta D_{MUX} - \Delta t_{PD}}{N} \tag{11.3}$$

Equation (11.3) implies that the oscillation period deviates from the ideal period (T_{REF}/N) to compensate the additional phase shift by the offset. That is, the output phase drifts during the VCO mode due to the frequency offset but is corrected when the reference edge is injected. As a result, the output phase periodically fluctuates at the reference frequency, while the peak-to-peak amplitude of the jitter equals to $\Delta t_{PD} + \Delta D_{MUX}$.

On the other hand, Figure 11.4(b) shows an example of a PLL with the phase detector offset, where we can see the difference over the MDLL. Because of the offset, the divided clock and the reference clock are locked with Δt_{PD} at the steady-state. It may lead to a reference spur as discussed in Chapter 6, but the fluctuation is relaxed very shortly (i.e., within PFD reset delay) so at least the oscillation frequency does not deviate from N/T_{REF}. As a result, since there is no phase drift, the output phase stays at a constant although there is a phase offset to the reference phase under locked. The observation above implies that the reference spur degradation caused by such mismatches is much severe for MDLLs. For quantifying the comparison, we can calculate the deterministic jitter of the PLL caused by the PD mismatch from (6.20), which can be rewritten to

$$\Delta\phi_{out} = \frac{2\omega_c}{\omega_{ref}} \cdot \Delta\phi_{err} = \frac{2\omega_c}{\omega_{ref}} \cdot \Delta t_{PD} \cdot \omega_{ref} \tag{11.4}$$

where $\Delta\phi_{out}$ is the output phase deviation introduced by the Δt_{PD}. Dividing by ω_{out}, we convert the output phase deviation to jitter as

$$\Delta t_{out} = \frac{\Delta\phi_{out}}{\omega_{out}} = \frac{2\omega_c}{\omega_{ref}} \cdot \Delta t_{PD} \cdot \frac{\omega_{ref}}{\omega_{out}} = \frac{2\omega_c}{N\omega_{ref}} \cdot \Delta t_{PD} \tag{11.5}$$

Because the reference spur is proportional to the deterministic jitter, the MDLL reference spur is normalized to the PLLs as

$$\frac{Spur_{MDLL}}{Spur_{PLL}} = \frac{\Delta t_{PD}}{\frac{2\omega_c}{N\omega_{ref}} \cdot \Delta t_{PD}} = \frac{N\omega_{ref}}{2\omega_c} \tag{11.6}$$

Considering that $\frac{\omega_{ref}}{\omega_c} > 10$, (11.6) means that the MDLL spur is typically ~40 dB worse than that of a PLL [6].

11.3 Offset compensation techniques

The phase offset is needed to be compensated since it introduces a severe deterministic jitter, as discussed in Section 11.2. Using a secondary loop where the phase offset is measured by a bang-bang PD has been proposed in [7]; however, the bang-bang PD also introduces another offset, so it may not be a complete solution. Reference [5] presented an offset-insensitive technique which is based on a high-resolution TDC and a subtraction with a correlator (similar to [4] of Chapter 10 and also refer to Figure 10.16). Reference [5] focused that the clock period where the reference edge is injected, T1, deviates from the rest of the periods (T0), by the

---

$$\frac{\Delta\phi_{out}}{\phi_{err}} \cong \frac{2\omega_c}{\omega_{ref}} = \frac{2f_c}{f_{ref}} \tag{6.20}$$

phase offset. To eliminate the offset, [5] shares the same TDC for measuring T1 and T0. The TDC output is stored in a register and then is compared to the adjacent one by a correlator. However, it is not an efficient solution because a high-resolution TDC consumes a huge power.

Based on the similar idea of comparing periods, [8–10] have proposed efficient offset calibration techniques. Rather than relying on a single loop in [5,8–10] use multiple feedback loops whose operations are divided in time, at the cost of increased lock time. Figure 11.5 shows the MDLL proposed in [8]. During the frequency calibration time, a pulse generator followed by a pulse DEMUX produces pulses whose widths equal to the periods of the MDLL output clock. The

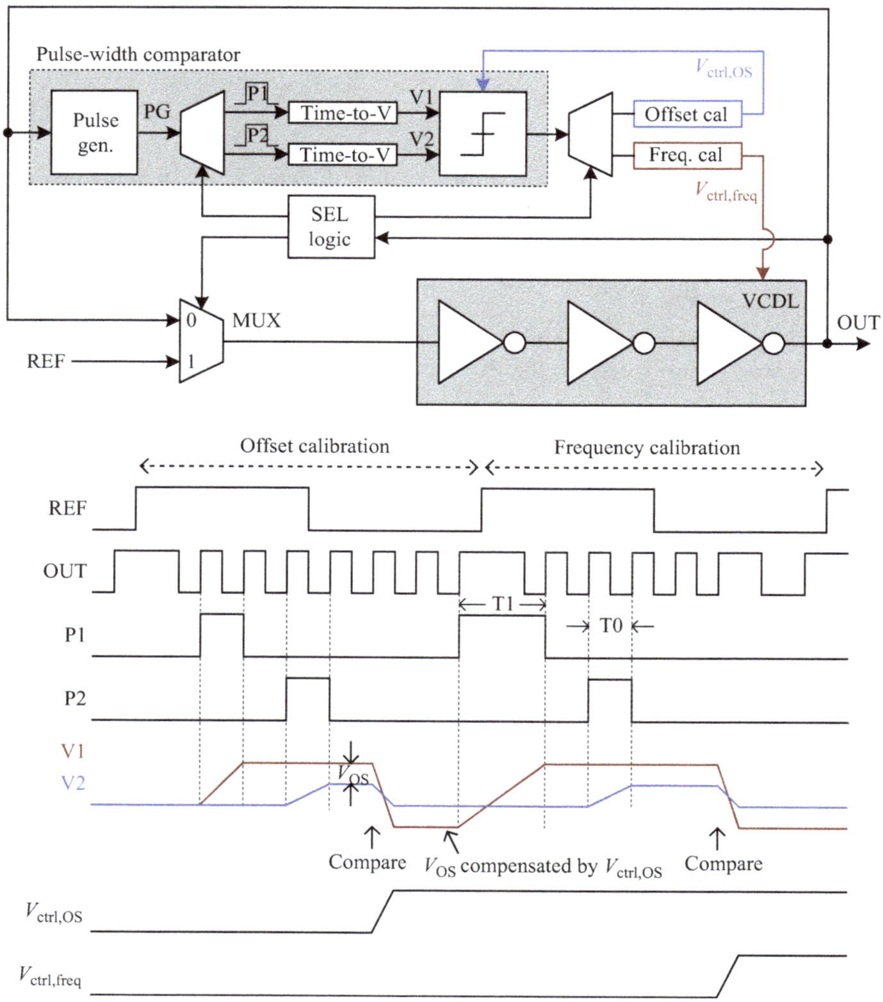

Figure 11.5 MDLL with time-division dual calibration, based on pulse-width comparator

pulse DEMUX passes two pulses, one of which is affected by the edge injection, to the next stage. Time-to-voltage converters produce two voltage levels which are proportional to the pulse-width of the incoming pulses. The following comparator decides that which voltage is higher, and the frequency calibration filter updates the VCDL control voltage ($V_{ctrl,freq}$) according to the decision. This frequency calibration process is similar to what is described in Section 10.5, and therefore any path mismatches, which can be introduced by MUX, DEMUX, time-to-voltage converters, and comparator, cause the static phase offset and the reference spur. Reference [8] has proposed to use an offset calibration loop that runs in background as well as the frequency tracking loop. During the offset calibration cycle, two periods which are not affected by the reference injection are selected for comparison. Ideally the resulting V1 and V2 are identical; however, the mismatches are converted to the voltage offset (V_{OS}). Then the comparator decides the polarity of the offset and feeds the decision to the offset calibration filter, where the collected offset information is accumulated. Since the frequency loop and offset loop run alternately and in the background, the MDLL is tolerable to voltage and temperature drift. In [9], a replica MUX and a bang-bang PD replace the relatively complex pulse-width comparator. Reference [10] presents an all-digital adjacent edge selector which extracts the periods of the MDLL clock.

11.4 Fractional-N MDLL and ILO

Because an MDLL (and an ILO) replaces one of the running edges of VCO with a reference edge, the VCO should run in an integer-multiples of the reference frequency; otherwise, an injection does not happen in a correct position. An example can be found in Figure 11.6, where a reference frequency (F_{REF}) is injected to an ideal VCO running at $4.25{\cdot}F_{REF}$. Because of the fractional ratio 0.25, the phase difference of the *REF* and the VCO clock drifts by 0.25 UI per T_{REF}. Therefore, the injection can happen at a correct timing only once in four times. The problem here is that the incorrect three injections replace the correct edges to the incorrect edges. Therefore, the injection continuously delays the VCO phase and therefore decreases the output frequency to $4{\cdot}F_{REF}$. Recall the injection locking phenomenon that an oscillator locks to the injection frequency due to the periodic phase shift. It seems that an injection cannot be allowed in a fractional-N clock multiplier.

However, a rotational injection technique proposed in [11] enables a fractional-N MDLL or ILO. Instead of injecting to a fixed position, [11] has

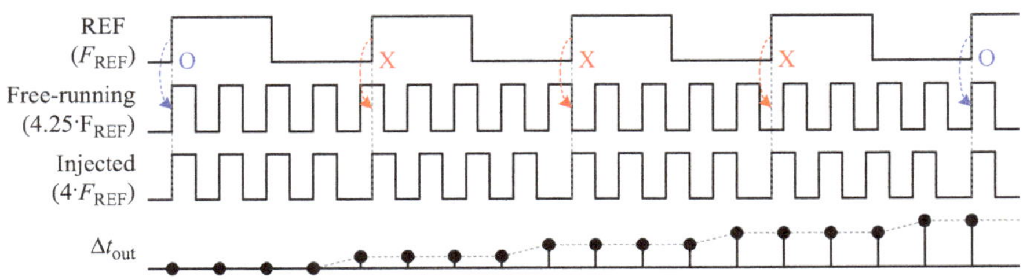

Figure 11.6 Reference injection to a fractional-N MDLL or ILO

proposed to rotate the injected delay element according to the fractional value. Since delay stages produce evenly spaced phases, rotating the delay element to be injected enables the fractional value up to the inverse of the number of the delay stages. For example, we looked at a 4.25 F_{REF} example in Figure 11.6, where there is a phase drift of 0.25 UI at every reference cycle. With four delay elements, each delay element has 0.25 UI delay. Therefore, the rotation lets the injection replace a correct edge as shown in Figure 11.7.

However, the fractional resolution of the rotational injection is bounded to the inverse of the number of delay stages. For a fine resolution, the number of stages needs to be increased, which degrades power consumption, phase noise, and speed. Instead, [12] has proposed to modulate the reference edge using a digitally-controlled delay line (DCDL), as shown in Figure 11.8. While placed at the front of

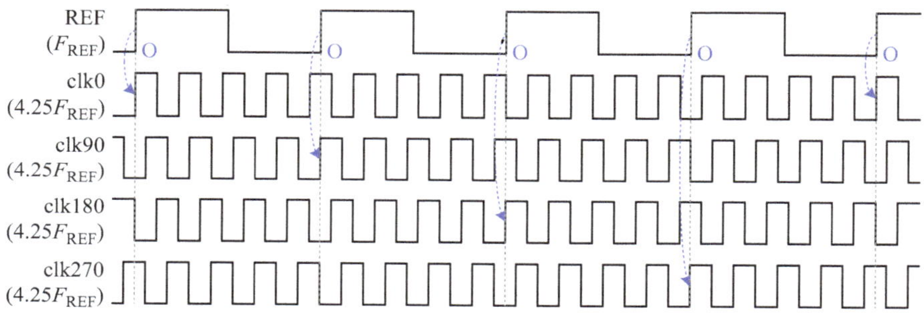

Figure 11.7 Fractional-N injection with rotational injection

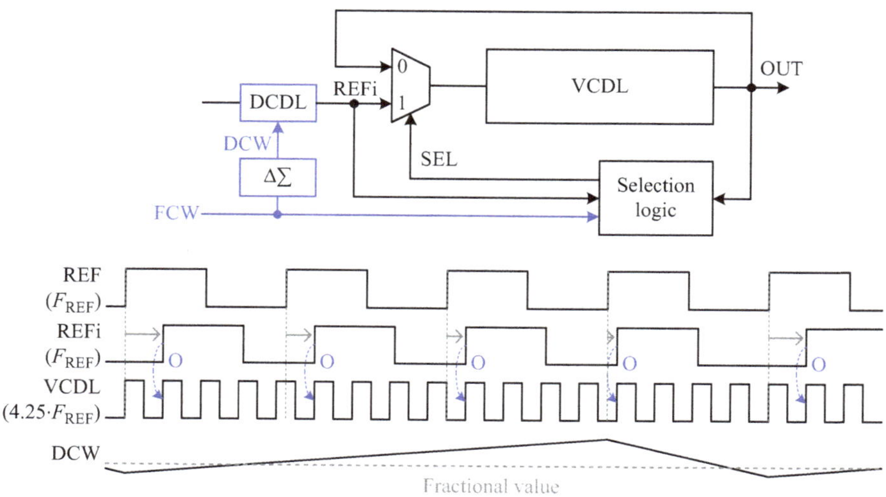

Figure 11.8 Fractional-N injection

the MUX, the DCDL adjusts the injection timing according to the delay control word (DCW). Therefore, a periodic modulation on the DCW lets the reference edge on a proper timing, while the edge is injected to a fixed delay element of the VCDL. The DCW modulation is simply achieved by accumulating the fractional part of the frequency control word (FCW). In this structure, the fractional resolution is not limited by the number of delay stages. However, a practical DCDL has finite resolution and nonlinearity between the DCW and the actual delay, which introduce output spur.

References

[1] Kim C, Hwang IC, Kang SM. A low-power small-area ±7.28-ps-jitter 1-GHz DLL-based clock generator. *IEEE Journal of Solid-State Circuits*. 2002;37(11):1414–1420.

[2] Farjad-Rad R, Dally W, Ng HT, *et al.* A low-power multiplying DLL for low-jitter multigigahertz clock generation in highly integrated digital chips. *IEEE Journal of Solid-State Circuits*. 2002;37(12):1804–1812.

[3] Ye S, Jansson L, Galton I. A multiple-crystal interface PLL with VCO realignment to reduce phase noise. *IEEE Journal of Solid-State Circuits*. 2002;37(12):1795–1803.

[4] Du Q, Zhuang J, Kwasniewski T. A low-phase noise, anti-harmonic programmable DLL frequency multiplier with period error compensation for spur reduction. *IEEE Transactions on Circuits and Systems II: Express Briefs*. 2006;53(11):1205–1209.

[5] Helal BM, Straayer MZ, Wei GY, Perrott MH. A highly digital MDLL-based clock multiplier that leverages a self-scrambling time-to-digital converter to achieve subpicosecond jitter performance. *IEEE Journal of Solid-State Circuits*. 2008;43(4):855–863.

[6] Gierkink SL. Low-spur, low-phase-noise clock multiplier based on a combination of PLL and recirculating DLL with dual-pulse ring oscillator and self-correcting charge pump. *IEEE Journal of Solid-State Circuits*. 2008; 43(12):2967–2976.

[7] Elshazly A, Inti R, Young B, Hanumolu PK. Clock multiplication techniques using digital multiplying delay-locked loops. *IEEE Journal of Solid-State Circuits*. 2013;48(6):1416–1428.

[8] Kim H, Kim Y, Kim T, Park H, Cho S. 19.3 A 2.4 GHz 1.5 mW digital MDLL using pulse-width comparator and double injection technique in 28 nm CMOS. In *2016 IEEE International Solid-State Circuits Conference (ISSCC)*. San Francisco, CA: IEEE; 2016 (pp. 328–329).

[9] Kim S, Ko HG, Cho SY, *et al.* 29.7 A 2.5 GHz injection-locked ADPLL with 197fs rms integrated jitter and −65dBc reference spur using time-division dual calibration. In *2017 IEEE International Solid-State Circuits Conference (ISSCC)*. San Francisco, CA: IEEE; 2017 (pp. 494–495).

[10] Yang S, Yin J, Mak PI, Martins RP. A 0.0056-mm 2–249-dB-FoM all-digital MDLL using a block-sharing offset-free frequency-tracking loop and dual multiplexed-ring VCOs. *IEEE Journal of Solid-State Circuits*. 2019; 54(1):88–98.

[11] Park P, Park J, Park H, Cho S. An all-digital clock generator using a fractionally injection-locked oscillator in 65nm CMOS. In *2012 IEEE International Solid-State Circuits Conference*. San Francisco, CA: IEEE; 2012 (pp. 336–337).

[12] Levantino S, Marucci G, Marzin G, Fenaroli A, Samori C, Lacaita AL. A 1.7 GHz fractional-N frequency synthesizer based on a multiplying delay-locked loop. *IEEE Journal of Solid-State Circuits*. 2015;50(11):2678–2691.

Appendix A
Figure of merits (FoMs) for evaluating VCOs and PLLs

VCOs and PLLs have many different specifications/applications and as well have several design trade-offs, such as between the jitter, power consumption, and frequency. As a result, a figure-of-merit (FoM), which is a benchmark used to characterize the performance of a device, needs to be defined to normalize different VCOs/PLLs performance for a fair comparison between them. Let us start with the FoM for VCO. Note that only white noise is under consideration, so we assume the phase noise is inversely proportional to the square of the offset frequency. Because it is very important to reflect the primary design trade-offs into the FoM, let us review the design trade-offs in a VCO and a PLL. From the theoretical phase noise of LC oscillator (4.8) and ring oscillator (4.18), we can first find common design trade-offs as

1. The phase noise is inversely proportional to the power consumption and
2. The phase noise is proportional to $\left(\frac{f_0}{\Delta f}\right)^2$.

Then we can simply express the VCO phase noise as

$$\mathscr{L}_{\mathrm{VCO}}(\Delta f) = \alpha \cdot \left(\frac{f_0}{\Delta f}\right)^2 \cdot \frac{1}{P_{\mathrm{VCO}}} \tag{A.1}$$

where f_0 and P_{VCO} represent the frequency and the power consumption of VCO. Note that the coefficient α is the only quality factor of an oscillator because the

---

$$\mathscr{L}(\Delta f) = 10\log\left(\frac{2kTF}{P_{sig}}\left(\frac{f_0}{2Q\Delta f}\right)^2 + \frac{2kT\gamma g_m R}{P_{sig}}\left(\frac{f_0}{2Q\Delta f}\right)^2\left(\frac{f_{1/f}}{\Delta f}\right)\right) \tag{4.8}$$

$$S_\phi(f) = \frac{8kT\gamma}{I(V_{DD} - V_{th})} \cdot \left(\frac{f_0}{f}\right)^2 \tag{4.18}$$

other terms just represent the trade-offs. As a result, we can find that the coefficient α can be a very good FoM. We can define the FoM for VCO as

$$\mathrm{FoM_{VCO}} = 10\log\left(\mathscr{L}_{\mathrm{VCO}}(\Delta f)\cdot\left(\frac{\Delta f}{f_0}\right)^2\cdot\left(\frac{P_{\mathrm{VCO}}}{1\,\mathrm{mW}}\right)\right) \tag{A.2}$$

Note that the power consumption is normalized to make the unit of FoM equal to that of phase noise, dBc/Hz.

As studied, the VCO noise dominates the out-band phase noise but the other PLL building blocks (PFD, charge pump, and divider) dominate the in-band phase noise (see Figure 6.12). So, we also need to see how the in-band noise is set and what the design trade-offs are. This Appendix will show the noise contribution by the charge-pump as an example. From the in-band transfer function and from the charge-pump to the output phase given in (8.3), we can obtain the output phase noise as

$$\frac{\Delta\phi_{\mathrm{out}}^2}{\Delta f} = \left(\frac{2\pi}{I_{\mathrm{CP}}}\cdot N\right)^2\cdot\frac{\Delta i_{\mathrm{CP}}^2}{\Delta f} \tag{A.3}$$

Considering that the I_{CP} flows only for the reset delay of the PFD every reference cycle at steady-state, the I_{CP} noise is expressed to

$$\frac{\Delta i_{\mathrm{CP}}^2}{\Delta f} = 4kT\gamma g_m\cdot\left(\frac{t_{\mathrm{PFD}}}{T_{\mathrm{ref}}}\right)^2 \tag{A.4}$$

By substituting (A.4) into (A.3), we obtain:

$$\begin{aligned}\frac{\Delta\phi_{\mathrm{out}}^2}{\Delta f} &= \left(16\pi^2 kT\gamma t_{\mathrm{PFD}}^2\cdot\frac{g_m}{I_{\mathrm{CP}}}\right)\cdot\frac{1}{I_{\mathrm{CP}}}\cdot\left(\frac{N}{T_{\mathrm{ref}}}\right)^2\\ &= \left(16\pi^2 kT\gamma t_{\mathrm{PFD}}^2\cdot\frac{2}{V_{\mathrm{OV}}}\right)\cdot\frac{1}{I_{\mathrm{CP}}}\cdot f_0^2\end{aligned} \tag{A.5}$$

where we can find that the in-band phase noise is inversely proportional to the power consumption (I_{CP}) but is proportional to the square of the output frequency. The phase noise contribution from other building blocks is derived in the elegant paper by Prof. Gao [1], so the readers can consult with if interested in. To summarize, [1] yields the same trade-off for the contribution from the others. As a result, we can write the in-band phase noise as

$$\mathscr{L}_{\mathrm{in-band}} = \beta\cdot f_0^2\cdot\frac{1}{P_{\mathrm{in-band}}}. \tag{A.6}$$

$$\frac{\Delta\phi_{\mathrm{out}}}{\Delta i_{\mathrm{PD,CP}}} \cong \frac{2\pi}{I_{\mathrm{CP}}}\cdot N \tag{8.3}$$

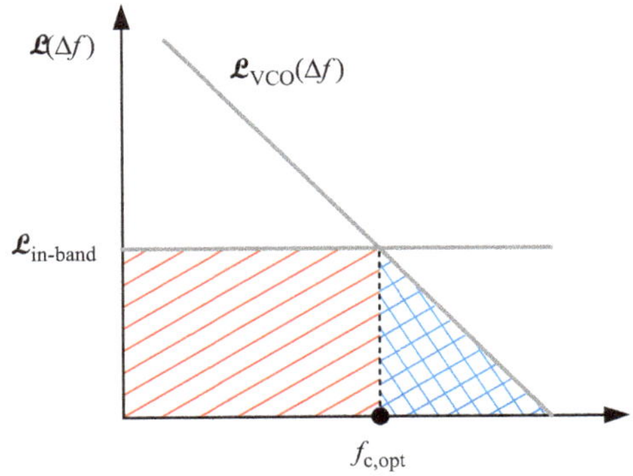

Figure A.1 PLL phase noise

Here we can repeat what we did for FoM_{VCO}, which leads to

$$\text{FoM}_{\text{in-band}} = \mathcal{L}_{\text{in-band}} \cdot P_{\text{in-band}} \cdot \frac{1}{f_0^2} \tag{A.7}$$

Let us go over to PLL phase noise. Unlike the VCO noise or the in-band noise, the PLL phase noise cannot be characterized at a single point because it is a combination of two straight lines as shown in Figure A.1. Instead, we need to integrate the entire phase noise plot to obtain the total output jitter as (recall (2.32))

$$\begin{aligned}
\sigma_{TIE}^2 &= \left(\frac{1}{2\pi f_0}\right)^2 \left(\int_0^{f_{c,opt}} \mathcal{L}_{\text{in-band}} d\Delta f + \int_{f_{c,opt}}^{\infty} \mathcal{L}_{VCO}(\Delta f) d\Delta f\right) \\
&= \left(\frac{1}{2\pi f_0}\right)^2 \left(\mathcal{L}_{\text{in-band}} \cdot f_{c,opt} + \int_{f_{c,opt}}^{\infty} \mathcal{L}_{VCO}(f_{c,opt}) \cdot \left(\frac{f_{c,opt}}{\Delta f}\right)^2 d\Delta f\right) \\
&= \left(\frac{1}{2\pi f_0}\right)^2 \left(\mathcal{L}_{\text{in-band}} \cdot f_{c,opt} + \mathcal{L}_{VCO}(f_{c,opt}) \cdot f_{c,opt}\right)
\end{aligned} \tag{A.8}$$

Note that we assumed that the PLL bandwidth is set to the optimum. Because $\mathcal{L}_{\text{in-band}}$ and $\mathcal{L}_{VCO}(\Delta f)$ intersect at $f_{c,opt}$, (A.8) is simplified to

$$\begin{aligned}
\sigma_{TIE}^2 &= \left(\frac{1}{2\pi f_0}\right)^2 \left(2 \cdot \mathcal{L}_{VCO}(f_{c,opt}) \cdot f_{c,opt}\right) \\
&= \left(\frac{1}{2\pi f_0}\right)^2 \left(2 \cdot \text{FoM}_{VCO} \cdot \left(\frac{f_0}{f_{c,opt}}\right)^2 \cdot \frac{1}{P_{VCO}} \cdot f_{c,opt}\right)
\end{aligned} \tag{A.9}$$

It also gives an interesting insight that, with the optimum bandwidth, the jitter contributions from the VCO noise and the in-band noise are the same. Equating $\mathscr{L}_{\text{in-band}}$ and $\mathscr{L}_{\text{VCO}}(f_{c,\text{opt}})$ yields:

$$\begin{aligned} \text{FoM}_{\text{in-band}} \cdot f_0^2 \cdot \frac{1}{P_{\text{in-band}}} &= \text{FoM}_{\text{VCO}} \cdot \left(\frac{f_0}{f_{c,\text{opt}}}\right)^2 \cdot \frac{1}{P_{\text{VCO}}} \\ f_{c,\text{opt}} &= \sqrt{\frac{\text{FoM}_{\text{VCO}} \cdot P_{\text{in-band}}}{\text{FoM}_{\text{in-band}} \cdot P_{\text{VCO}}}} \end{aligned} \tag{A.10}$$

Substituting (A.10) into (A.9) leads to

$$\sigma_{\text{TIE}}^2 = \left(\frac{1}{2\pi}\right)^2 \left(2 \cdot \sqrt{\frac{\text{FoM}_{\text{in-band}} \cdot \text{FoM}_{\text{VCO}}}{P_{\text{in-band}} \cdot P_{\text{VCO}}}}\right). \tag{A.11}$$

Based on the inequality of arithmetic and geometric means, (A.11) is minimized when

$$P_{\text{in-band}} = P_{\text{VCO}} = \frac{1}{2} P_{\text{PLL}} \tag{A.12}$$

Using (A.12), (A.11) is simplified to

$$\sigma_{\text{TIE}}^2 = \left(\frac{1}{\pi}\right)^2 \cdot \frac{\sqrt{\text{FoM}_{\text{in-band}} \cdot \text{FoM}_{\text{VCO}}}}{P_{\text{PLL}}}, \tag{A.13}$$

where we can find that there is a trade-off between the total integrated jitter and the power consumption, which is very important in PLL design. Again, we can use the coefficient as FoM so we can define the PLL FoM as

$$\text{FoM}_{\text{PLL}} = 10 \log\left(\left(\frac{\sigma_{\text{TIE}}}{1\text{ s}}\right)^2 \cdot \left(\frac{P_{\text{PLL}}}{1\text{ mW}}\right)\right). \tag{A.14}$$

Note that the jitter and the power are normalized to 1 s and 1 mW, respectively, so the unit of the FoM is dB.

On the other hand, (A.14) does not consider the contribution from the reference clock noise. Since the reference phase noise is amplified by 20 logN, the high division factor causes a significant degradation on the total jitter, in case that the reference noise is the dominant source. For such a case, we can define another FoM which includes the division factor:

$$\text{FoM}_{\text{PLL},N} = 10 \log\left(\left(\frac{\sigma_{\text{TIE}}}{1\text{ s}}\right)^2 \cdot \left(\frac{P_{\text{PLL}}}{1\text{ mW}}\right) \cdot \frac{1}{N}\right) \tag{A.15}$$

$$\sigma_{\text{TIE}} = \frac{1}{2\pi f_0}\sqrt{2 \cdot \int_{f_{\min}}^{f_{\max}} 10^{\mathscr{L}(\Delta f)} df} \tag{2.32}$$

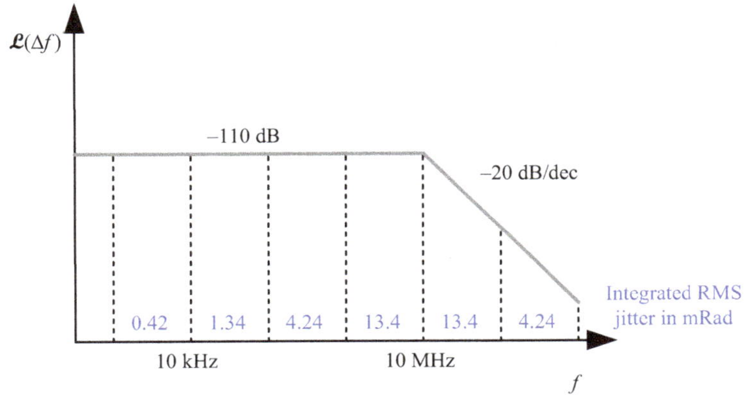

Figure A.2 Example of PLL jitter integration by decade

(A.15) may be more effective to evaluate IL-PLLs and MDLLs, where the output phase noise is dominated by the reference noise since the jitter tracking bandwidth is much wider than that of the conventional PLLs.

Let us briefly see how the FoM of a practical PLL is characterized. The measurement of power consumption is very straightforward, but there are many possible ways for the measurement of jitter. One is to use of oscilloscope's eye diagram. However, it highly depends on which trigger source is used and how long the eye diagram is accumulated. For example, if the reference clock of a PLL is used as a trigger source, the jitter contribution from the reference clock will not be fully considered. As a result, typically the jitter is measured by integrating the phase noise by using a spectrum analyzer. What is important here is the integration range. Figure A.2 shows an example of a PLL phase noise. The numbers written in blue are the integrated jitter in mRad integrated over a decade. Starting from the low frequency, we can find that the amount of the integrated jitter exponentially increases and hits the maximum at the vicinity of the cut-off frequency. But it starts decreasing exponentially after the cut-off frequency. That is, the total integrated jitter is dominated by that at the vicinity of the cut-off frequency. It becomes even larger when we consider jitter peaking. As a result, the integration range is typically 3~4 decades, wherein the cut-off frequency should be included.

Reference

[1] Gao X, Klumperink EA, Geraedts PF, Nauta B. Jitter analysis and a benchmarking figure-of-merit for phase-locked loops. *IEEE Transactions on Circuits and Systems II: Express Briefs*. 2009;56(2):117–121.

Appendix B

Survey on state-of-the-art clock generators

In this appendix, a brief survey on the state-of-the-art clock generators is presented. Figure B.1 shows the FoMs of CMOS clock generators presented in the International Solid-State Circuits Conference (ISSCC) from 2010 to 2019 [1–91]. In the graph, each class is distinguished with different colors, shapes, and fillings. First, colors are used to distinguish the clock generation techniques. The ADPLLs, SS-PLLs, IL-PLLs, and MDLLs are marked with red, green, blue, and black, respectively. Others are marked with purple, which are mainly analog charge-pump PLLs. Second, the clock generators based on LC oscillators are specified with a circle while ring oscillator is marked with a triangle. Third, the integer-N and the fractional-N are also distinguished whether it is filled or not. One important observation is that, more than 80% of the state-of-the-art clock generators rely on ADPLL, SS-PLL, IL-PLL, or MDLL techniques, as shown in the breakdown (Figure B.2). That is the primary reason why this book covers those four techniques through Chapters 8–11. We can also find that the FoM has continued to improve, thanks to many researchers' dedications and efforts. The trend of the improvement is around −10 dB every 4 years. The readers may prejudge that a part of the improvement has benefited from the CMOS process scaling. However, as we can observe in Figure B.3, there is no strong correlation between the FoM and the process technology nodes. Until most recently, the 65 nm CMOS technology is the most popular process (47 out of 92 designs) among the state-of-the-art clock generators. It implies that this kind of analog circuits that deal with the noise may not benefit much from the process scaling. The scaling does enhance the power efficiency of the building blocks owing to the reduction of supply voltage and intrinsic capacitance; however, increased 1/f noise and mismatch degrade the noise performance. The reduced supply voltage also lowers the signal amplitude, which degrades the SNR. We may infer that the upside and downside of the process scaling on the clocking circuit are balanced until now, so that explains why there is no strong correlation between the performance and the process node. On the other hand, there is a trend that the silicon area shrinks as the process scales. Figure B.4 shows the trend. We can observe that the area of a clock of the state-of-the-art clock generators follow the scaling except for the designs of 65 nm process node.

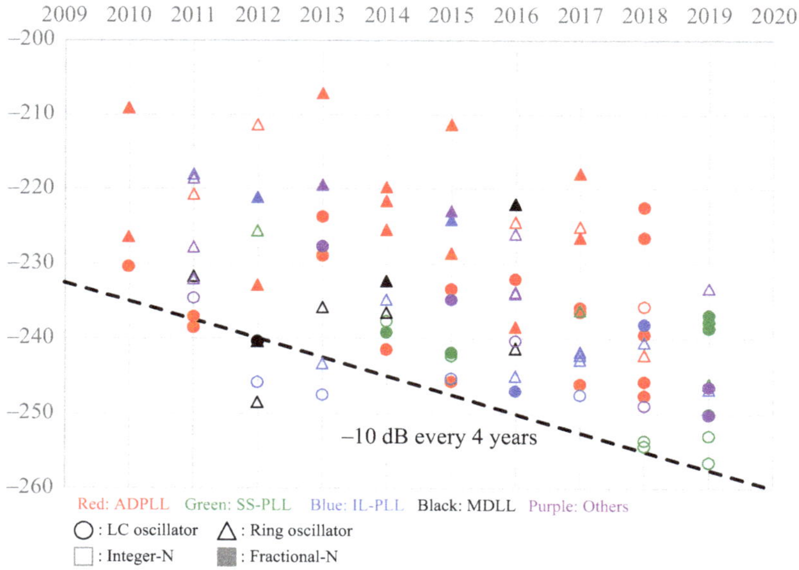

Figure B.1 FoMs of CMOS clock generators presented in ISSCC 2010–2019

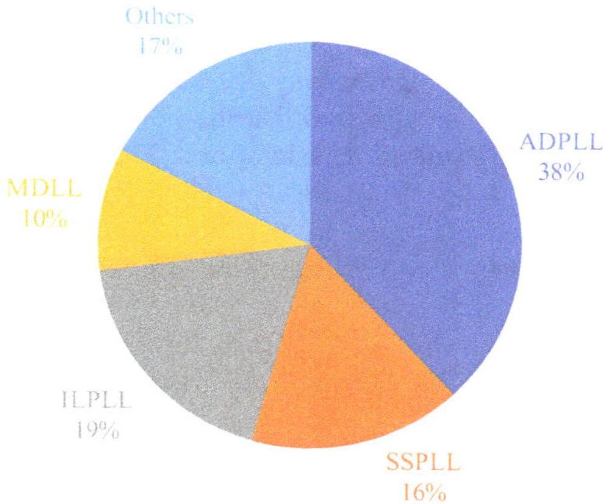

Figure B.2 Breakdown of clock generation techniques in ISSCC 2010–2019

Considering that more than 50% efforts have been focused on the 65 nm node, we may conclude that the process scaling at least guarantees the reduction in area consumption. Moreover, we can also find that the ADPLLs are leading the area scaling because they take full advantage of the process scaling.

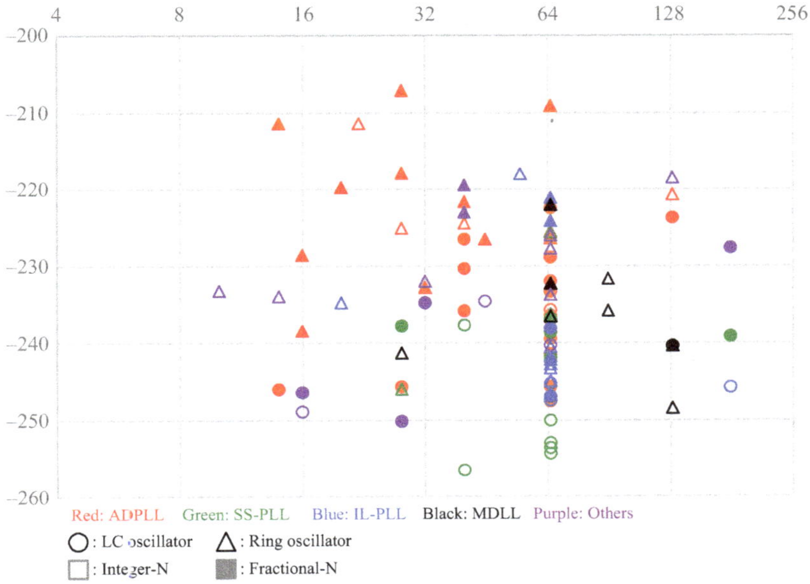

Figure B.3 FoM trends with respect to the CMOS process technology

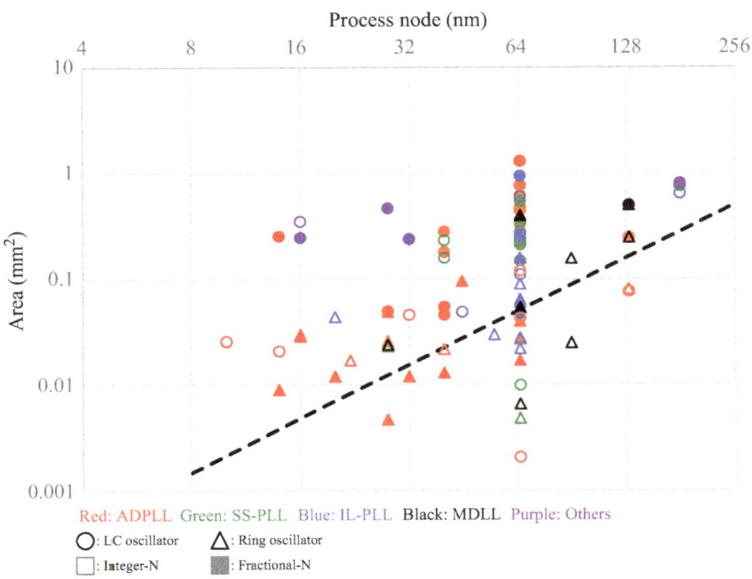

Figure B.4 Silicon area versus process node

In addition to the scaling of the silicon area, the scaling also helps increasing the clock frequency especially for the ring-based clock generators as we discussed before, that is, ring oscillators benefit from the scaled process. Figure B.5, where the operating frequencies of ring-based clock generators are plotted with respect to the process technology nodes, evidences that it shows ~30% speed improvement per scaling of one generation (channel length scaled by $1/\sqrt{2}$).

Silicon area occupancy of a clock generation circuit is another important metric since it determines the fabrication cost; however, it is not considered in the conventional FoM. We would expect that a clock generator exhibits a better performance if it consumes more area since it enables using robust but bulky circuit elements (inductor and capacitor) and including more circuit functionality. This hypothesis is partially verified in Figure B.6, where the FoM is plotted with respect to the silicon area. In Figure B.6, we can see that there is –10 dB/decade trend line which implies the FoM is inversely proportional to the area. We can also find that the ring-based and the LC-based generators stand on the same line. It shows the area-noise trade-off between the ring and LC oscillators, which partially justifies introducing an area-included FoM to compare the ring-based and LC-based clock generators.

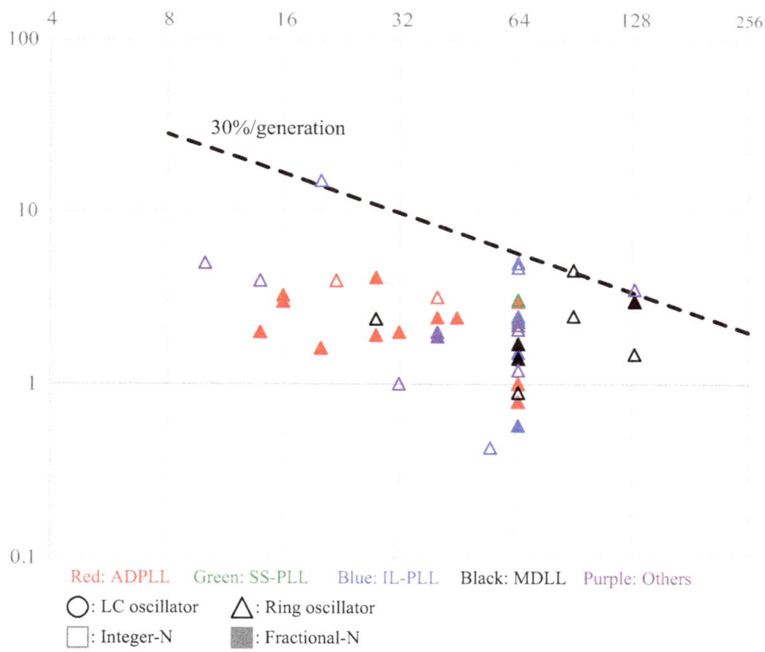

Figure B.5 Operating frequency of ring-based clock generator versus process node

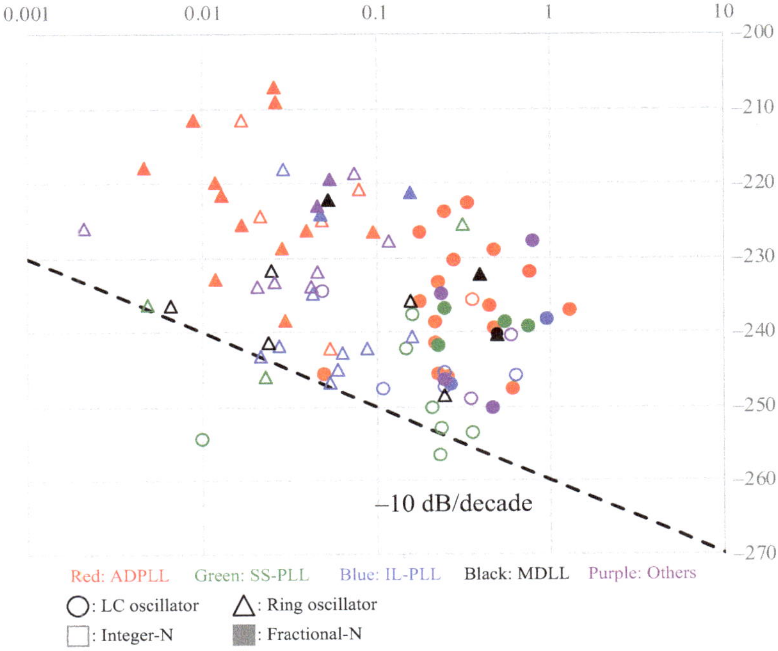

Figure B.6 FoM trends with respect to silicon area

On the other hand, we need to investigate the correlation between the FoM and the division factor N. It can be explained in two different ways. At first, we studied that the input reference noise is amplified by 20 logN at a low-frequency region (in-band). Since the reference noise level is not a matter of the integrated circuit technology, it has not been improved rapidly as the silicon technology. As a result, if the phase noise from the integrated circuits becomes better and better, the reference noise will eventually dominate the in-band phase noise, so we can say that the overall phase noise will be constrained by the reference phase noise. We can observe that trend in both ring-based clock generators and LC-based clock generators. The FoMs of the ring-based generators and the LC-based generators are separately drawn in Figure B.7 and Figure B.8, respectively, because the reference clock holds different amounts of share for the phase noise of them. Nevertheless, the FoM degrades by ~10 dB/decade for both of the rings and the LCs. Such trend becomes more evident when only the IL-PLLs and MDLLs are focused, due to their inherent high dependency on the reference jitter multiplication. This observation justifies the use of FoM_N given in (A.15) for high-end clock generators rather than the conventional FoM, especially for IL-PLLs and MDLLs.

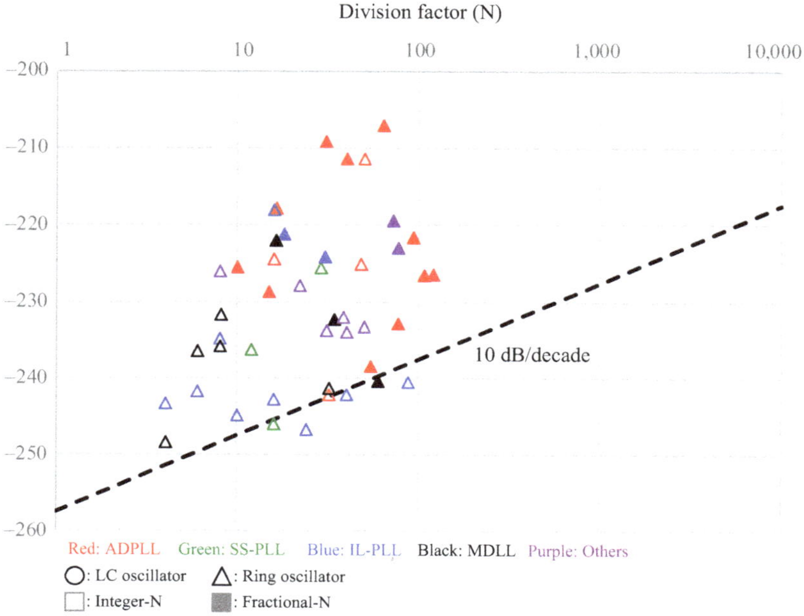

Figure B.7 FoM of ring-based clock generator versus division factor

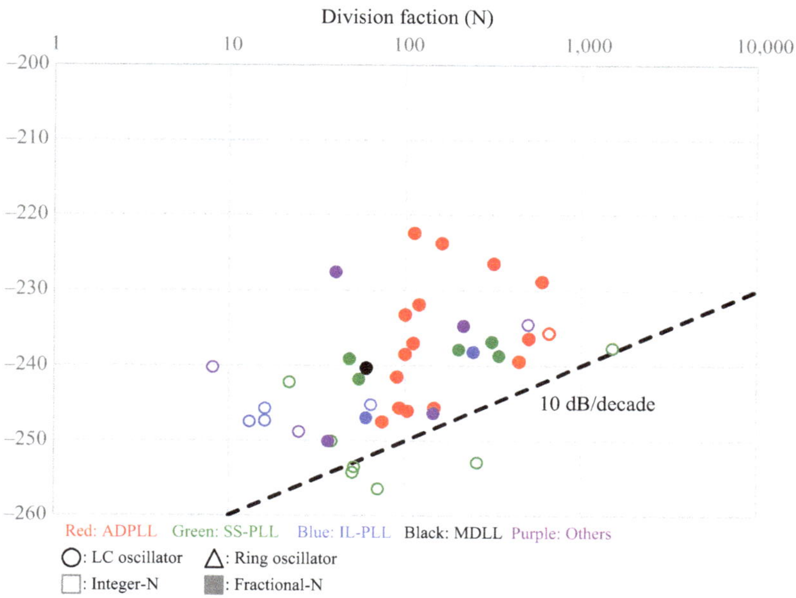

Figure B.8 FoM of LC-based clock generator versus division factor

The dependency on the reference clock frequency is also examined. Because a higher reference frequency is highly correlated to a lower division factor, we can expect that it has a similar trend as that with the division factor. Here we can see the second way of explanation. If we assume that the reference clock is clean enough, the noise from integrated circuits dominates the overall phase noise. Note that the VCO phase noise is usually the dominant one. In order to suppress the VCO noise, a PLL needs a higher loop bandwidth, which is the motivation of the IL-PLL and MDLL. Recall that bandwidth of a conventional PLL is limited to 1/10 of the reference frequency, and that of the IL-PLL and MDLL is limited to 1/2 of the reference frequency. That means a higher reference frequency filters out more VCO phase noise. The survey results shown in Figures B.9 and B.10 verify our hypothesis.

After observing such dependency on the division factor and the reference frequency, the readers may wonder if we can observe the same improvement over time for FoM_N like Figure B.11. Fortunately, we can observe that the CMOS clock generators have been improved by the same –10 dB/4 years in terms of FoM_N as well. From a different point of view, it can be another justification of using FoM_N for characterizing and comparing clock generators.

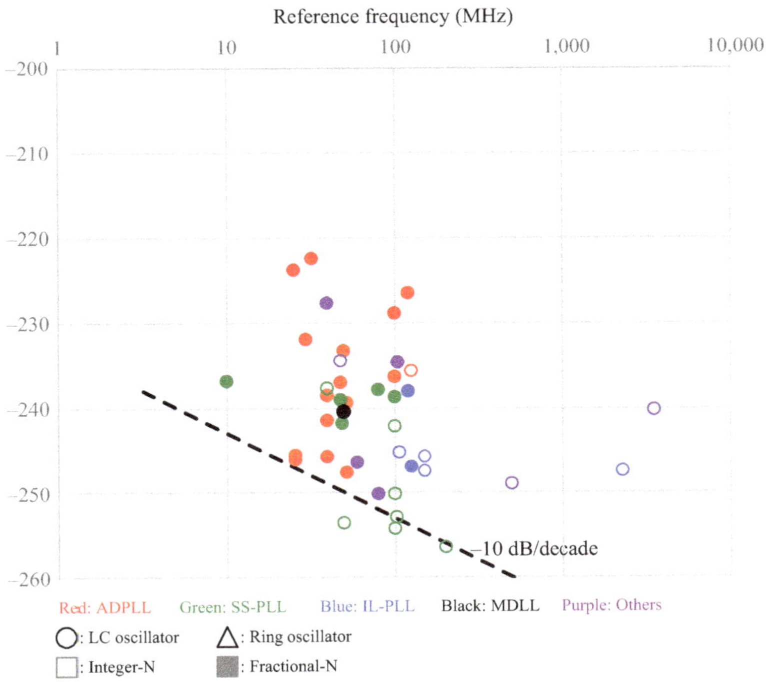

Figure B.9 FoM of ring-based clock generator versus reference clock frequency

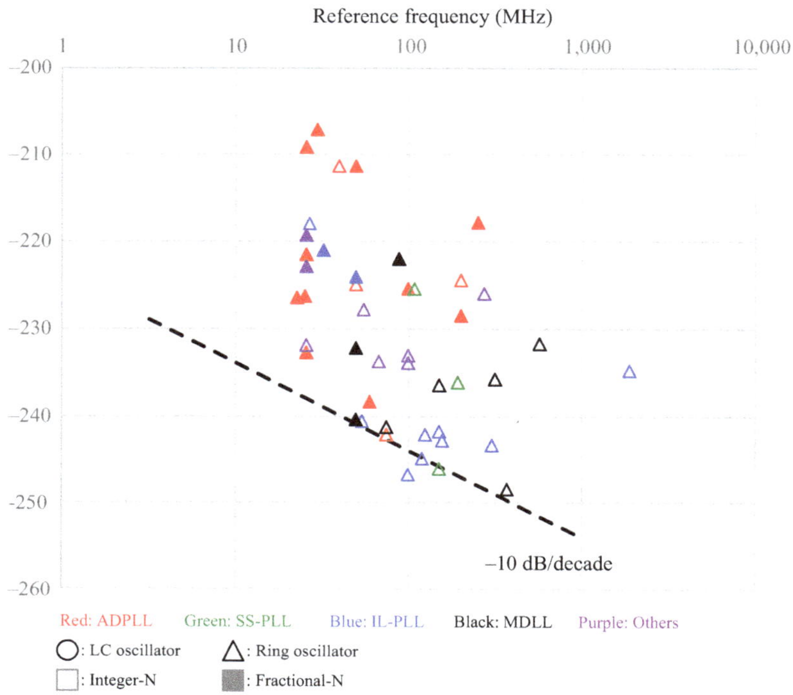

Figure B.10 FoM of LC-based clock generator versus reference clock frequency

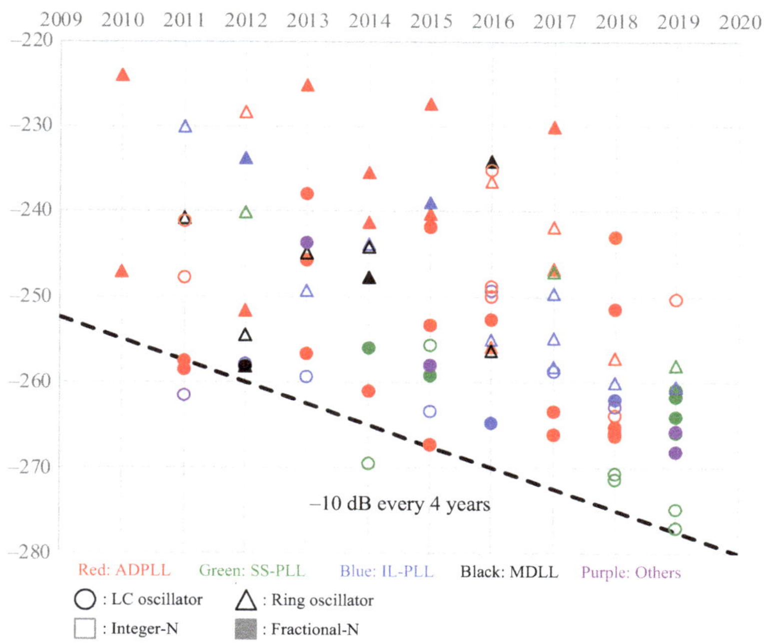

Figure B.11 FoMN of CMOS clock generators presented in ISSCC 2010–2019

References

[1] Huang Z., Luong H. C. "An 82-to-108 GHz −181 dB-FOMTADPLL employing a DCO with split-transformer and dual-path switched-capacitor ladder and a clock-skew-sampling delta-sigma TDC." in *IEEE International Solid-State Circuits Conference Digest of Technical Papers (ISSCC)*, February 2018, pp. 260–262.

[2] Liu H., Tang D., Sun Z., *et al.* "A 0.98 mW fractional-N ADPLL using 10b isolated constant-slope DTC with FOM of −246 dB for IoT applications in 65 nm CMOS." in *IEEE International Solid-State Circuits Conference Digest of Technical Papers (ISSCC)*, February 2018, pp. 246–248.

[3] Cherniak D., Grimaldi L., Bertulessi L., Samori C., Nonis R., Levantino S. "A 23 GHz low-phase-noise digital bang-bang PLL for fast triangular and saw-tooth chirp modulation." in *IEEE International Solid-State Circuits Conference Digest of Technical Papers (ISSCC)*, February 2018, pp. 248–250.

[4] Weyer D., Dayanik M. B., Jang S., Flynn M. P. "A 36.3-to-38.2 GHz −216 dBc/Hz 240 nm CMOS fractional-N FMCW chirp synthesizer PLL with a continuous-time bandpass delta-sigma time-to-digital converter." in *IEEE International Solid-State Circuits Conference Digest of Technical Papers (ISSCC)*, February 2018, pp. 250–252.

[5] Bertulessi L., Grimaldi L., Cherniak D., Samori C., Levantino S. "A low-phase-noise digital bang-bang PLL with fast lock over a wide lock range." in *IEEE International Solid-State Circuits Conference Digest of Technical Papers (ISSCC)*, February 2018, pp. 252–254.

[6] Ho C., Chen M. S., "A fractional-N digital PLL with background-dither-noise-cancellation loop achieving <−62.5 dBc worst-case near-carrier fractional spurs in 65 nm CMOS." in *IEEE International Solid-State Circuits Conference Digest of Technical Papers (ISSCC)*, February 2018, pp. 394–396.

[7] Hussein A., Vasadi S., Soliman M., Paramesh J. "19.3 A 50-to-66 GHz 65 nm CMOS all-digital fractional-N PLL with 220 fs rms jitter." in *IEEE International Solid-State Circuits Conference Digest of Technical Papers (ISSCC)*, February 2017, pp. 326–327.

[8] He Y., Liu Y. H., Kuramochi T., *et al.* "24.7 A 673 μW 1.8-to-2.5 GHz dividerless fractional-N digital PLL with an inherent frequency-capture capability and a phase-dithering spur mitigation for IoT applications." in *IEEE International Solid-State Circuits Conference Digest of Technical Papers (ISSCC)*, February 2017, pp. 420–421.

[9] Yao C. W., Ni R., Lau C., *et al.* "24.8 A 14 nm fractional-N digital PLL with 0.14 ps rms jitter and −78 dBc fractional spur for cellular RFICs." in *IEEE International Solid-State Circuits Conference Digest of Technical Papers (ISSCC)*, February 2017, pp. 422–423.

[10] Ho C., Chen M. S. "10.5 A digital PLL with feedforward multi-tone spur cancelation loop achieving <−73 dBc fractional spur and <−110 dBc Reference Spur in 65 nm CMOS." in *IEEE International Solid-State Circuits Conference Digest of Technical Papers (ISSCC)*, February 2016, pp. 190–191.

[11] Gao X., Tee L., Wu W., *et al.* "9.4 A 28 nm CMOS digital fractional-N PLL with −245.5 dB FOM and a frequency tripler for 802.11abgn/ac radio." in *IEEE International Solid-State Circuits Conference Digest of Technical Papers (ISSCC)*, February 2015, pp. 1–3.

[12] Kim H., Sang J., Jo Y., *et al.* "14.4 A 5 GHz −95 dBc-reference-Spur 9.5 mW digital fractional-N PLL using reference-multiplied time-to-digital converter and reference-spur cancellation in 65 nm CMOS." in *IEEE International Solid-State Circuits Conference Digest of Technical Papers (ISSCC)*, February 2015, pp. 1–3.

[13] Marzin G., Levantino S., Samori C., Lacaita A. L. "2.9 A background calibration technique to control bandwidth in digital PLLs." in *IEEE International Solid-State Circuits Conference Digest of Technical Papers (ISSCC)*, February 2014, pp. 54–55.

[14] Wu W., Bai X., Staszewski R. B., Long J. R. "A 56.4-to-63.4 GHz spurious-free all-digital fractional-N PLL in 65 nm CMOS." in *IEEE International Solid-State Circuits Conference Digest of Technical Papers (ISSCC)*, February 2013, pp. 352–353.

[15] Nonis R., Grollitsch W., Santa T., Cherniak D., Dalt N. D. "A 2.4 ps rms-jitter digital PLL with multi-output bang-bang phase detector and phase-interpolator-based fractional-N divider." in *IEEE International Solid-State Circuits Conference Digest of Technical Papers (ISSCC)*, February 2013, pp. 356–357.

[16] Pavlovic N., Bergervoet J. "A 5.3 GHz digital-to-time-converter-based fractional-N all-digital PLL." in *IEEE International Solid-State Circuits Conference Digest of Technical Papers (ISSCC)*, February 2011, pp. 54–56.

[17] Tasca D., Zanuso M., Marzin G., Levantino S., Samori C., Lacaita A. L. "A 2.9-to-4.0 GHz fractional-N digital PLL with bang-bang phase detector and 560 fs rms integrated jitter at 4.5 mW power." in *IEEE International Solid-State Circuits Conference Digest of Technical Papers (ISSCC)*, February 2011, pp. 88–90.

[18] Borremans J., Vengattaramane K., Giannini V., Craninckx J. "A 86 MHz-to-12 GHz digital-intensive phase-modulated fractional-N PLL using a 15pJ/Shot 5 ps TDC in 40 nm digital CMOS." in *IEEE International Solid-State Circuits Conference Digest of Technical Papers (ISSCC)*, February 2010, pp. 480–481.

[19] Seong T., Lee Y., Yoo S., Choi J. "A −242 dB FOM and −75 dBc-reference-spur ring-DCO-based all-digital PLL using a fast phase-error correction technique and a low-power optimal-threshold TDC." in *IEEE International Solid-State Circuits Conference Digest of Technical Papers (ISSCC)*, February 2018, pp. 396–398.

[20] Jang T., Jeong S., Jeon D., Choo K. D., Sylvester D., Blaauw D. "8.4 A 2.5 ps 0.8-to-3.2 GHz bang-bang phase- and frequency-detector-based all-digital PLL with noise self-adjustment." in *IEEE International Solid-State Circuits Conference Digest of Technical Papers (ISSCC)*, February 2017, pp. 148–149.

[21] Yeh C., Hsieh C., Liu S. "19.5 A 3.2 GHz digital phase-locked loop with background supply-noise cancellation." in *IEEE International Solid-State Circuits Conference Digest of Technical Papers (ISSCC)*, February 2016, pp. 332–333.

[22] August N., Lee H., Vandepas M., Parker R. "A TDC-less ADPLL with 200-to-3200 MHz range and 3 mW power dissipation for mobile SoC clocking in 22 nm CMOS." in *IEEE International Solid-State Circuits Conference Digest of Technical Papers (ISSCC)*, February 2012, pp. 246–248.

[23] Elshazly A., Inti R., Yin W., Young B., Hanumolu P. K. "A 0.4-to-3 GHz digital PLL with supply-noise cancellation using deterministic background calibration." in *IEEE International Solid-State Circuits Conference Digest of Technical Papers (ISSCC)*, February 2011, pp. 92–94.

[24] Cho H., Seong K., Cho K. H., *et al.* "8.7 A 0.0047 mm 2 highly synthesizable TDC- and DCO-less fractional-N PLL with a seamless lock range of fREF to 1 GHz." in *IEEE International Solid-State Circuits Conference Digest of Technical Papers (ISSCC)*, February 2017, pp. 154–155.

[25] Kong L., Razavi B. "19.5 A 2.4 GHz RF fractional-N synthesizer with 0.25fREFBW." in *IEEE International Solid-State Circuits Conference Digest of Technical Papers (ISSCC)*, February 2017, pp. 330–331.

[26] Ahmad Z., Lee M., Kenneth K. O. "20.5 1.4THz, −13 dBm-EIRP frequency multiplier chain using symmetric- and asymmetric-CV varactors in 65 nm CMOS." in *IEEE International Solid-State Circuits Conference Digest of Technical Papers (ISSCC)*, February 2016, pp. 350–351.

[27] Tsai T., Yuan M., Chang C., Liao C., Li C., Staszewski R. B. "14.5 A 1.22 ps integrated-jitter 0.25-to-4 GHz fractional-N ADPLL in 16 nm FinFET CM0S." in *IEEE International Solid-State Circuits Conference Digest of Technical Papers (ISSCC)*, February 2015, pp. 1–3.

[28] Song M., Kim T., Kim J., Kim W., Kim S., Park H. "14.8 A 0.009 mm 22.06 mW 32-to-2000 MHz 2nd-order ΔΣ analogous bang-bang digital PLL with feed-forward delay-locked and phase-locked operations in 14 nm FinFET technology." in *IEEE International Solid-State Circuits Conference Digest of Technical Papers (ISSCC)*, February 2015, pp. 1–3.

[29] Liu J., Jang T.-K., Lee Y., *et al.* "15.2 A 0.012 mm^2 3.1 mW bang-bang digital fractional-N PLL with a power-supply-noise cancellation technique and a walking-one-phase-selection fractional frequency divider." in *IEEE International Solid-State Circuits Conference Digest of Technical Papers (ISSCC)*, February 2014, pp. 268–269.

[30] Huang Y., Liang C., Huang H., Wang P. "15.3 A 2.4 GHz ADPLL with digital-regulated supply-noise-insensitive and temperature-self-compensated ring DCO." in *IEEE International Solid-State Circuits Conference Digest of Technical Papers (ISSCC)*, February 2014, pp. 270–271.

[31] Elkholy A., Elshazly A., Saxena S., Shu G., Hanumolu P. K. "15.4 A 20-to-1000 MHz ±14 ps peak-to-peak jitter reconfigurable multi-output all-digital clock generator using open-loop fractional dividers in 65 nm CMOS." in

IEEE International Solid-State Circuits Conference Digest of Technical Papers (ISSCC), February 2014, pp. 272–273.

[32] Jang T. K., Nan X., Liu F., *et al.* "A 0.026 mm 25.3 mW 32-to-2000 MHz digital fractional-N phase locked-loop using a phase-interpolating phase-to-digital converter." in *IEEE International Solid-State Circuits Conference Digest of Technical Papers (ISSCC)*, February 2013, pp. 254–255.

[33] Hong J. P., Kim S. J., Liu J., *et al.* "A 0.004 mm 2250 μW ΔΣ TDC with time-difference accumulator and a 0.012 mm 22.5 mW bang-bang digital PLL using PRNG for low-power SoC applications." in *IEEE International Solid-State Circuits Conference Digest of Technical Papers (ISSCC)*, February 2012, pp. 240–242.

[34] Chen M. S., Su D., Mehta S. "A calibration-free 800 MHz fractional-N digital PLL with embedded TDC." in *IEEE International Solid-State Circuits Conference Digest of Technical Papers (ISSCC)*, February 2012, pp. 472–473.

[35] Grollitsch W., Nonis R., Da Dalt N. "A 1.4 ps rms-period-jitter TDC-less fractional-N digital PLL with digitally controlled ring oscillator in 65 nm CMOS." in *IEEE International Solid-State Circuits Conference Digest of Technical Papers (ISSCC)*, February 2012, pp. 478–479.

[36] Kim J., Yoon H., Lim Y., *et al.* "16.2 A 76 fs rms jitter and –40 dBc integrated-phase-noise 28-to-31 GHz frequency synthesizer based on digital sub-sampling PLL using optimally spaced voltage comparators and background loop-gain optimization." in *IEEE International Solid-State Circuits Conference Digest of Technical Papers (ISSCC)*, February 2019, pp. 258–260.

[37] Yang Z., Chen Y., Yang S., Mak P., Martins R. P. "16.8 A 25.4-to-29.5 GHz 10.2 mW isolated sub-sampling PLL achieving −252.9 dB jitter-power FoM and −63 dBc reference spur." in *IEEE International Solid-State Circuits Conference Digest of Technical Papers (ISSCC)*, February 2019, pp. 270–272.

[38] Zhang Z., Zhu G., Yue C. P. "30.8 A 0.65V 12-to-16 GHz sub-sampling PLL with 56.4 fs rms integrated jitter and −256.4 dB FoM." in *IEEE International Solid-State Circuits Conference Digest of Technical Papers (ISSCC)*, February 2019, pp. 488–490.

[39] Sharkia A., Mirabbasi S., Shekhar S. "A 0.01 mm 24.6-to-5.6 GHz sub-sampling type-I frequency synthesizer with −254 dB FOM." in *IEEE International Solid-State Circuits Conference Digest of Technical Papers (ISSCC)*, February 2018, pp. 256–258.

[40] Sharma J., Krishnaswamy H. "A dividerless reference-sampling RF PLL with −253.5 dB jitter FOM and <−67 dBc reference spurs." in *IEEE International Solid-State Circuits Conference Digest of Technical Papers (ISSCC)*, February 2018, pp. 258–260.

[41] Siriburanon T., Kondo S., Kimura K., *et al.* "25.2 A 2.2 GHz −242 dB-FOM 4.2 mW ADC-PLL using digital sub-sampling architecture." in *IEEE International Solid-State Circuits Conference Digest of Technical Papers (ISSCC)*, February 2015, pp. 1–3.

[42] Szortyka V., Shi Q., Raczkowski K., Parvais B., Kuijk M., Wambacq P. "21.4 A 42 mW 230fs-jitter sub-sampling 60 GHz PLL in 40 nm CMOS." in *IEEE International Solid-State Circuits Conference Digest of Technical Papers (ISSCC)*, February 2014, pp. 366–367.

[43] Liu H., Sun Z., Huang H., *et al.* "16.1 A 265 μW fractional-N Digital PLL with seamless automatic switching subsampling/sampling feedback path and duty-cycled frequency-locked loop in 65 nm CMOS." in *IEEE International Solid-State Circuits Conference Digest of Technical Papers (ISSCC)*, February 2019, pp. 256–258.

[44] Grimaldi L., Bertulessi L., Karman S., *et al.* "16.7 A 30 GHz digital sub-sampling fractional-N PLL with 198 fs rms jitter in 65 nm LP CMOS." in *IEEE International Solid-State Circuits Conference Digest of Technical Papers (ISSCC)*, February 2019, pp. 268–270.

[45] Shi Q., Bunsen K., Markulic N., Craninckx J. "26.1 A self-calibrated 16 GHz subsampling-PLL-based 30 s fast chirp FMCW modulator with 1.5 GHz bandwidth and 100kHz rms error." in *IEEE International Solid-State Circuits Conference Digest of Technical Papers (ISSCC)*, February 2019, pp. 408–410.

[46] Chen Z. Z., Wang Y. H., Shin J., *et al.* "14.9 sub-sampling all-digital fractional-N frequency synthesizer with −111 dBc/Hz in-band phase noise and an FOM of −242 dB." in *IEEE International Solid-State Circuits Conference Digest of Technical Papers (ISSCC)*, February 2015, pp. 1–3.

[47] Huang P., Chang W., Lee T. "21.2 A 2.3 GHz fractional-N dividerless phase-locked loop with −112 dBc/Hz in-band phase noise." in *IEEE International Solid-State Circuits Conference Digest of Technical Papers (ISSCC)*, February 2014, pp. 362–363.

[48] Yang X., Chan C., Zhu Y., Martins R. P. "16.3 A −246 dB jitter-FoM 2.4 GHz calibration-free ring-oscillator PLL achieving 9% jitter variation over PVT." in *IEEE International Solid-State Circuits Conference Digest of Technical Papers (ISSCC)*, February 2019, pp. 260–262.

[49] Chuang J., Krishnaswamy H. "19.4 A 0.0049 mm 22.3 GHz sub-sampling ring-oscillator PLL with time-based loop filter achieving −236.2 dB jitter-FOM." in *IEEE International Solid-State Circuits Conference Digest of Technical Papers (ISSCC)*, February 2017, pp. 328–329.

[50] Sai A., Kobayashi Y., Saigusa S., Watanabe O., Itakura T. "A digitally stabilized type-III PLL using ring VCO with 1.01 ps rms integrated jitter in 65 nm CMOS." in *IEEE International Solid-State Circuits Conference Digest of Technical Papers (ISSCC)*, February 2012, pp. 248–250.

[51] Yoo S., Choi S., Kim J., Yoon H., Lee Y., Choi J. "19.2 A PVT-robust −39 dBc 1kHz-to-100 MHz integrated-phase-noise 29 GHz injection-locked frequency multiplier with a 600μW frequency-tracking loop using the averages of phase deviations for mm-band 5G transceivers." in *IEEE International Solid-State Circuits Conference Digest of Technical Papers (ISSCC)*, February 2017, pp. 324–325.

[52] Elkholy A., Talegaonkar M., Anand T., Hanumolu P. K. "10.7 A 6.75-to-8.25 GHz 2.25 mW 190 fs rms integrated-jitter PVT-insensitive injection-

locked clock multiplier using all-digital continuous frequency-tracking loop in 65 nm CMOS." in *IEEE International Solid-State Circuits Conference Digest of Technical Papers (ISSCC)*, February 2015, pp. 1–3.

[53] Lee I., Chen Y., Liu S., Jou C., Hsueh F., Hsieh H. "A divider-less sub-harmonically injection-locked PLL with self-adjusted injection timing." in *IEEE International Solid-State Circuits Conference Digest of Technical Papers (ISSCC)*, February 2013, pp. 414–415.

[54] Huang Y., Liu S. "A 2.4 GHz sub-harmonically injection-locked PLL with self-calibrated injection timing." in *IEEE International Solid-State Circuits Conference Digest of Technical Papers (ISSCC)*, February 2012, pp. 338–340.

[55] Yoon H., Kim J., Park S., *et al.* "A −31 dBc integrated-phase-noise 29 GHz fractional-N frequency synthesizer supporting multiple frequency bands for backward-compatible 5G using a frequency doubler and injection-locked frequency multipliers." in *IEEE International Solid-State Circuits Conference Digest of Technical Papers (ISSCC)*, February 2018, pp. 366–368.

[56] Elkholy A., Elmallah A., Elzeftawi M., Chang K., Hanumolu P. K. "10.6 A 6.75-to-8.25 GHz, 250 fs rms-integrated-jitter 3.25 mW rapid on/off PVT-insensitive fractional-N injection-locked clock multiplier in 65 nm CMOS." in *IEEE International Solid-State Circuits Conference Digest of Technical Papers (ISSCC)*, February 2016, pp. 192–193.

[57] Yoo S., Choi S., Lee Y., Seong T., Lim Y., Choi J. "30.9 A 140 fs rms-jitter and -72 dBc-reference-spur ring-VCO-based injection-locked clock multiplier using a background triple-point frequency/phase/slope calibrator." in *IEEE International Solid-State Circuits Conference Digest of Technical Papers (ISSCC)*, February 2019, pp. 490–492.

[58] Megawer K. M., Elkholy A., Coombs D., Ahmed M. G., Elmallah A., Hanumolu P. K. "A 5 GHz 370 fs rms 6.5 mW clock multiplier using a crystal-oscillator frequency quadrupler in 65 nm CMOS." in *IEEE International Solid-State Circuits Conference Digest of Technical Papers (ISSCC)*, February 2019, pp. 392–394.

[59] Ngo H. C., Nakata K., Yoshioka T., Terashima Y., Okada K., Matsuzawa A. "8.5 A 0.42 ps-jitter −241.7 dB-FOM synthesizable injection-locked PLL with noise-isolation LDO." in *IEEE International Solid-State Circuits Conference Digest of Technical Papers (ISSCC)*, February 2017, pp. 150–151.

[60] Coombs D., Elkholy A., Nandwana R. K., Elmallah A., Hanumolu P. K. "8.6 A 2.5-to-5.75 GHz 5 mW 0.3 ps rms-jitter cascaded ring-based digital injection-locked clock multiplier in 65 nm CMOS." in *IEEE International Solid-State Circuits Conference Digest of Technical Papers (ISSCC)*, February 2017, pp. 152–153.

[61] Kim S., Ko H. G., Cho S. Y., *et al.* "29.7 A 2.5 GHz injection-locked ADPLL with 197 fs rms integrated jitter and −65 dBc reference spur using time-division dual calibration." in *IEEE International Solid-State Circuits Conference Digest of Technical Papers (ISSCC)*, February 2017, pp. 494–495.

[62] Choi S., Yoo S., Choi J. "10.7 A 185 fs rms-integrated-jitter and −245 dB FOM PVT-robust ring-VCO-based injection-locked clock multiplier with a

continuous frequency-tracking loop using a replica-delay cell and a dual-edge phase detector." in *IEEE International Solid-State Circuits Conference Digest of Technical Papers (ISSCC)*, February 2016, pp. 194–195.

[63] Chien J. C., Upadhyaya P., Jung H., *et al.* "2.8 A pulse-position-modulation phase-noise-reduction technique for a 2-to-16 GHz injection-locked ring oscillator in 20 nm CMOS." in *IEEE International Solid-State Circuits Conference Digest of Technical Papers (ISSCC)*, February 2014, pp. 52–53.

[64] Deng W., Musa A., Siriburanon T., Miyahara M., Okada K., Matsuzawa A. "A 0.022 mm 2970 μW dual-loop injection-locked PLL with −243 dB FOM using synthesizable all-digital PVT calibration circuits." in *IEEE International Solid-State Circuits Conference Digest of Technical Papers (ISSCC)*, February 2013, pp. 248–249.

[65] Liang C., Hsiao K. "An injection-locked ring PLL with self-aligned injection window." in *IEEE International Solid-State Circuits Conference Digest of Technical Papers (ISSCC)*, February 2011, pp. 90–92.

[66] Deng W., Yang D., Narayanan A. T., *et al.* "14.1 A 0.048 mm^2 3 mW synthesizable fractional-N PLL with a soft injection-locking technique." in *IEEE International Solid-State Circuits Conference Digest of Technical Papers (ISSCC)*, February 2015, pp. 1–3.

[67] Park P., Park J., Park H., Cho S. "An all-digital clock generator using a fractionally injection-locked oscillator in 65 nm CMOS." in *IEEE International Solid-State Circuits Conference Digest of Technical Papers (ISSCC)*, February 2012, pp. 336–337.

[68] Park D., Cho S. "A 14.2 mW 2.55-to-3 GHz cascaded PLL with reference injection, 800 MHz delta-sigma modulator and 255 fs rms integrated jitter in 0.13 μm CMOS." in *IEEE International Solid-State Circuits Conference Digest of Technical Papers (ISSCC)*, February 2012, pp. 344–346.

[69] Kim H., Kim Y., Kim T., Park H., Cho S. "19.3 A 2.4 GHz 1.5 mW digital MDLL using pulse-width comparator and double injection technique in 28 nm CMOS." in *IEEE International Solid-State Circuits Conference Digest of Technical Papers (ISSCC)*, February 2016, pp. 328–329.

[70] Deng W., Yang D., Ueno T., *et al.* "15.1 A 0.0066 mm 2 780 μW fully synthesizable PLL with a current-output DAC and an interpolative phase-coupled oscillator using edge-injection technique." in *IEEE International Solid-State Circuits Conference Digest of Technical Papers (ISSCC)*, February 2014, pp. 266–267.

[71] Anand T., Talegaonkar M., Elshazly A., Young B., Hanumolu P. K. "A 2.5 GHz 2.2 mW/25 μW on/off-state power 2 ps rms-long-term-jitter digital clock multiplier with 3-reference-cycles power-on time." in *IEEE International Solid-State Circuits Conference Digest of Technical Papers (ISSCC)*, February 2013, pp. 256–257.

[72] Elshazly A., Inti R., Young B., Hanumolu P. K. "A 1.5 GHz 890 μW digital MDLL with 400 fs rms integrated jitter, −55.6 dBc reference spur and 20 fs/mV supply-noise sensitivity using 1b TDC." in *IEEE International Solid-*

State Circuits Conference Digest of Technical Papers (ISSCC), February 2012, pp. 242–244.

[73] Ali T. A., Hafez A. A., Drost R., Ho R., Yang C. K. "A 4.6 GHz MDLL with −46 dBc reference spur and aperture position tuning." in *IEEE International Solid-State Circuits Conference Digest of Technical Papers (ISSCC)*, February 2011, pp. 466–468.

[74] Kundu S., Kim B., Kim C. H. "19.2 A 0.2-to-1.45 GHz subsampling fractional-N all-digital MDLL with zero-offset aperture PD-based spur cancellation and in-situ timing mismatch detection." in *IEEE International Solid-State Circuits Conference Digest of Technical Papers (ISSCC)*, February 2016, pp. 326–327.

[75] Marucci G., Fenaroli A., Marzin G., Levantino S., Samori C., Lacaita A. L. "21.1 A 1.7 GHz MDLL-based fractional-N frequency synthesizer with 1.4 ps RMS integrated jitter and 3 mW power using a 1b TDC." in *IEEE International Solid-State Circuits Conference Digest of Technical Papers (ISSCC)*, February 2014, pp. 360–361.

[76] Turker D., Bekele A., Upadhyaya P., *et al.* "A 7.4-to-14 GHz PLL with 54 fs rms jitter in 16 nm FinFET for integrated RF-data-converter SoCs." in *IEEE International Solid-State Circuits Conference Digest of Technical Papers (ISSCC)*, February 2018, pp. 378–380.

[77] Agrawal A., Natarajan A. "2.2 A scalable 28 GHz coupled-PLL in 65 nm CMOS with single-wire synchronization for large-scale 5G mm-wave arrays." in *IEEE International Solid-State Circuits Conference Digest of Technical Papers (ISSCC)*, February 2016, pp. 38–39.

[78] Osorio J. F., Vaucher C. S., Huff B., Heijden E. V. D., de Graauw A. "A 21.7-to-27.8 GHz 2.6-degrees-rms 40 Mw frequency synthesizer in 45 nm CMOS for mm-Wave communication applications." in *IEEE International Solid-State Circuits Conference Digest of Technical Papers (ISSCC)*, February 2011, pp. 278–280.

[79] Song F., Zhao Y., Wu B., Tang L., Lin L., Razavi B. "16.5 A fractional-N synthesizer with 110 fs rms jitter and a reference quadrupler for wideband 802.11ax." in *IEEE International Solid-State Circuits Conference Digest of Technical Papers (ISSCC)*, February 2019, pp. 264–266.

[80] Yang D., Abidi A., Darabi H., *et al.* "16.6 A calibration-free triple-loop bang-bang PLL achieving 131 fs rms jitter and-70 dBc fractional spurs." in *IEEE International Solid-State Circuits Conference Digest of Technical Papers (ISSCC)*, February 2019, pp. 266–268.

[81] Ferriss M., Sadhu B., Rylyakov A., Ainspan H., Friedman D. "10.9 A 13.1-to-28 GHz fractional-N PLL in 32 nm SOI CMOS with a ΔΣ noise-cancellation scheme." in *IEEE International Solid-State Circuits Conference Digest of Technical Papers (ISSCC)*, February 2015, pp. 1–3.

[82] Tang Z., Wan X., Wang M., Liu J. "A 50-to-930 MHz quadrature-output fractional-N frequency synthesizer with 770-to-1860 MHz single-inductor LC-VCO and without noise folding effect for multistandard DTV tuners." in

IEEE International Solid-State Circuits Conference Digest of Technical Papers (ISSCC), February 2013, pp. 358–359.

[83] Fan Y., Xiang B., Zhang D., Ayers J. S., Shen K. J., Mezhiba A. "19.5 digital leakage compensation for a low-power and low-jitter 0.5-to-5 GHz PLL in 10 nm FinFET CMOS Technology." in *IEEE International Solid-State Circuits Conference Digest of Technical Papers (ISSCC)*, February 2019, pp. 320–322.

[84] Huang Z., Jiang B., Li L., Luong H. C. "2.3 A 4.2 μs-settling-time 3rd-order 2.1 GHz phase-noise-rejection PLL using a cascaded time-amplified clock-skew sub-sampling DLL." in *IEEE International Solid-State Circuits Conference Digest of Technical Papers (ISSCC)*, February 2016, pp. 40–41.

[85] Shen K. Y., Farooq S. F., Fan Y., *et al.* "19.4 A 0.17-to-3.5 mW 0.15-to-5 GHz SoC PLL with 15 dB built-in supply noise rejection and self-bandwidth control in 14 nm CMOS." in *IEEE International Solid-State Circuits Conference Digest of Technical Papers (ISSCC)*, February 2016, pp. 330–331.

[86] Zhu J., Nandwana R. K., Shu G., Elkholy A., Kim S., Hanumolu P. K. "19.8 A 0.0021 mm 21.82 mW 2.2 GHz PLL using time-based integral control in 65 nm CMOS." in *IEEE International Solid-State Circuits Conference Digest of Technical Papers (ISSCC)*, February 2016, pp. 338–340.

[87] Lee H., Kern A. M., Hyvonen S., Young I. A. "A scalable sub-1.2 mW 300 MHz-to-1.5 GHz host-clock PLL for system-on-chip in 32 nm CMOS." in *IEEE International Solid-State Circuits Conference Digest of Technical Papers (ISSCC)*, February 2011, pp. 96–97.

[88] Sai A., Yamaji T., Itakura T. "A 570 fs rms integrated-jitter ring-VCO-based 1.21 GHz PLL with hybrid loop." in *IEEE International Solid-State Circuits Conference Digest of Technical Papers (ISSCC)*, February 2011, pp. 98–100.

[89] Hwang S., Song M., Kwak Y., Jung I., Kim C. "A 0.076 mm 23.5 GHz spread-spectrum clock generator with memoryless Newton–Raphson modulation profile in 0.13 μm CMOS." in *IEEE International Solid-State Circuits Conference Digest of Technical Papers (ISSCC)*, February 2011, pp. 360–362.

[90] Liang C., Wang P. "10.8 A wideband fractional-N ring PLL using a near-ground pre-distorted switched-capacitor loop filter." in *IEEE International Solid-State Circuits Conference Digest of Technical Papers (ISSCC)*, February 2015, pp. 1–3.

[91] Kao T., Liang C., Chiu H., Ashburn M. "A wideband fractional-N ring PLL with fractional-spur suppression using spectrally shaped segmentation." in *IEEE International Solid-State Circuits Conference Digest of Technical Papers (ISSCC)*, February 2013, pp. 416–417.

Appendix C

System Verilog modeling of CMOS clock generator including jitter

In order to design a big system, it is very important to make a reasonable model of an analog circuit, as it is used to explore how the performance of the analog circuit affects the system performance, without running a time-consuming simulation with physical devices or gates. System Verilog is widely used in those fields for high-level simulations including analog circuit models. In this appendix, an example of System Verilog modeling of CMOS clock generator is provided. Based on the phase noise and jitter sources of a generic clock generator that we studied in Chapter 6, we include the following jitters in the model: (1) the white (random) jitter induced from the white noise of the output buffer stage, (2) the oscillator phase noise, (3) the phase noise of reference clock, and (4) the sinusoidal jitter. Basically, the purpose of the clock jitter model is to get the phase noise profile from a designer and translate it to the actual timing perturbation.

At first, the model defines an ideal clock without jitter. And the jittery clock sources are defined by modulating the edge timing of the ideal clock. The white jitter is relatively easily modeled in System Verilog by using the built-in Gaussian process function. Given the Gaussian distribution function, we can adjust the standard deviation (or amplitude) of the Gaussian, which is the RMS value of the white jitter. The RMS jitter (TIE, time-interval error) is obtained as

$$\sigma_{\text{rand}}^2 = \left(\frac{1}{2\pi f_0}\right)^2 \int_0^{f_{\text{int}}} 10^{\mathcal{L}_{\text{white}}} df = \left(\frac{1}{2\pi f_0}\right)^2 \cdot f_{\text{int}} \cdot 10^{\mathcal{L}_{\text{white}}} \tag{C.1}$$

where f_0, f_{int}, and $\mathcal{L}(\Delta f)$ represent the clock frequency, the frequency integration range, and the phase noise of the white noise floor. We can express f_{int} as a function of f_0 as

$$f_{\text{int}} = a \cdot f_{\text{int}} \tag{C.2}$$

where a is a constant, which can be used as a fitting parameter in the model for better accuracy. Combining (C.1) and (C.2) leads to

$$\sigma_{\text{rand}}^2 = a \cdot \left(\frac{1}{2\pi f_0}\right)^2 \cdot f_0 \cdot 10^{\mathcal{L}_{\text{white}}} \tag{C.3}$$

Then the period of the clock with the white jitter is expressed as

$$T_{\text{rand}} = T_{\text{ideal}} + \sigma_{\text{rand}} \cdot \$\text{dist\_normal} \tag{C.4}$$

The input reference phase noise is also modeled based on (C.1) and (C.4). The sinusoidal jitter can be easily added on to (C.4), resulting in

$$T_{\text{rand}} = T_{\text{ideal}} + \sigma_{\text{rand}} \cdot \$\text{dist}_{\text{normal}} + A_{\sin} T_{\text{ideal}} \sin(2\pi f_0 t) \tag{C.5}$$

where $A_{\sin}$ is the amplitude of the sinusoidal modulation in UI.

Modeling of oscillator noise is a little bit tricky because it is a Gaussian process. Recalling the phase noise of ring oscillator, the white noise from the buffer stage induces a white jitter at the time of the noise injection (similar to the white jitter from the output buffer); however, the feedback of the oscillator accumulates the jitter permanently. As a result, we can find that the oscillator noise is a Wiener process, which is an integration of the white noise. An important observation here is that the period jitter does not include such accumulating effect, whereas the TIE does. It means the period jitter can be modeled using the Gaussian process function, and the TIE is obtained by accumulating the period jitter. Note that an accumulator can be easily implemented in System Verilog (i.e., $y[n] <= y[n\text{-}1] + x[n]$). The RMS amplitude of the period jitter is obtained from the formula derived in [1]:

$$\sigma_{\text{period}}^2 = \left(\frac{1}{2\pi f_0}\right)^2 \cdot 8 \cdot \int_0^{\infty} 10^{\mathcal{L}(f)} \sin^2(\pi f T_{clk}) df \tag{C.6}$$

In this particular case, (C.6) is simplified to

$$\begin{aligned} \sigma_{\text{period}}^2 &= \left(\frac{1}{2\pi f_0}\right)^2 \cdot 8 \cdot 10^{\mathcal{L}_{\text{white,VCO}}} \cdot \int_0^{f_{\text{int}}} \sin^2(\pi f T_{\text{ideal}}) df \\ &= \left(\frac{1}{2\pi f_0}\right)^2 \cdot 8 \cdot 10^{\mathcal{L}_{\text{white,VCO}}} \cdot \int_0^{f_{\text{int}}} \frac{1 - \cos(2\pi f T_{\text{ideal}})}{2} df \end{aligned} \tag{C.7}$$

Assuming that the integration of the cosine term gives a negligible value, (C.7) is further simplified to

$$\sigma_{\text{period}}^2 = \left(\frac{1}{\pi f_0}\right)^2 \cdot 10^{\mathcal{L}_{\text{white,VCO}}} \cdot f_{\text{int}} = \left(\frac{1}{\pi f_0}\right)^2 \cdot 10^{\mathcal{L}_{\text{white,VCO}}} \cdot a \cdot f_0 \tag{C.8}$$

Since $\mathcal{L}_{\text{white,VCO}}$ will not be given from the phase noise profile of a clock but $\mathcal{L}_{\text{white}}$ will be given, we can replace $\mathcal{L}_{\text{white,VCO}}$ by introducing another fitting coefficient k as

$$\sigma^2_{\text{period}} = \left(\frac{1}{\pi f_0}\right)^2 \cdot 10^{\mathcal{L}_{\text{white}}} \cdot k \cdot a \cdot f_0 \tag{C.9}$$

On the other hand, in the phase noise profile, the oscillator phase noise is fully defined with the white noise level and the white noise crossover frequency (f_c). That is, even though the $\mathcal{L}_{\text{white}}$ is the same, the oscillator phase noise is proportional to the crossover frequency. As a result, we can express an effective period jitter fed to the accumulator as

$$\sigma^2_{\text{period,eff}} = \left(\frac{1}{\pi f_0}\right)^2 \cdot b \cdot 10^{\mathcal{L}_{\text{white}}} \cdot f_0 \cdot f_c \tag{C.10}$$

Recalling the jitter transfer function of the clock generation circuits (PLL, MDLL, ILPLL, ...), they show a generic filtering nature such that the reference jitter is low-pass filtered but the oscillator jitter is high-pass filtered, and the cut-off frequencies are the same. A first-order low-pass filter can be easily implemented in System Verilog as

$$y[n] = x[n] \cdot \alpha + y[n-1] \cdot (1-\alpha) \tag{C.11}$$

In (C.11), the coefficient α determines the cut-off frequency of the filter. (C.11) is rewritten as

$$x[n] - y[n] = \frac{1-\alpha}{\alpha} \cdot (y[n] - y[n-1]) \tag{C.12}$$

Comparing (C.12) with a generic expression of a continuous-time low-pass filter of

$$\text{in}(t) - \text{out}(t) = \frac{1}{\omega_c} \cdot \frac{d\text{out}(t)}{dt} \tag{C.13}$$

leads to

$$\omega_c = \frac{\alpha}{T_s(1-\alpha)} \tag{C.14}$$

or

$$\alpha = \frac{\omega_c}{1/T_s + \omega_c} \tag{C.15}$$

With (C.15), we can build a low-pass filter whose cut-off frequency equals to the given cut-off frequency from the designer. The high-pass filter for the oscillator jitter is implemented in a similar way as

$$y[n] = (x[n] - x[n-1]) \cdot (1 - \alpha) + y[n-1] \cdot (1 - \alpha) \tag{C.16}$$

In the actual model, two identical filters are cascaded to realize the second-order filtering of a practical clock generation circuit. The complete System Verilog model is given in Listing C.1. Figure C.1 shows the simulated phase noise profiles from the model. The three profiles are the phase noise with optimum JTB, high JTB, and low JTB, respectively. Each of the profiles includes the phase noise profiles of the input reference, the oscillator, and the output buffer. The results match well with the theoretically expected ones given in Figure 6.12.

Listing C.1. System Verilog model of clock generator with various sources of jitter.

```
// Jitter Clock Generator Model
// author: Woorham Bae
'timescale 1ns/1fs
module clkgen(
input var real sine_jtt_frq, // Sine jitter frequency(GHz)
input var real sine_jtt_mag, // Sine jitter amplitude(UI)
input var real ref_pn, // reference PN level(dBc/Hz)
input var real white_pn, // White PN level(dBc/Hz)
input var real white_cutoff, // White PN cut-off freq(Hz)
input var real pll_bw, // PLL loop BW freq(Hz)
output reg clk_ideal, // Ideal clock
output reg clk_jtt_PM, // PM (Buffer) Jitter clock
output reg clk_jtt_FM, // FM (VCO) Jitter clock
output reg clk_jtt_PMFM, // PM+FM Jitter clock
output reg clk_jtt_PLL // PM+FM+ref Jitter filtered by PLL clock
);
import "DPI" function real sin(real x);
parameter real f_clk = 5.0; // Clock frequency (GHz)
real p_clk_ideal = 1/f_clk; // Ideal clock period
real par_pll_bw = pll_bw/1e9; // PLL BW translated into digital filter
real p_clk_jtt_rand = 1*p_clk_ideal; // PM jitter clock period
real p_clk_jtt_in = 1*p_clk_ideal; // Input jitter clock period
real p_clk_jtt_acc = 10*p_clk_ideal; // VCO jitter clock period + default delay to prevent negative delay
real p_clk_jtt_acc_flt_1st = 0.0; // VCO jitter filtered by 1st order
real p_clk_jtt_acc_flt = 0.0; // VCO jitter filtered by 2nd order
real p_clk_jtt_acc_flt_d = 0.0; // For implementing filter
real p_clk_jtt_in_flt_1st = 0.0; // Input jitter filtered by 1st order
real p_clk_jtt_in_flt = 0.0; // Input jitter filtered by 2nd order
```

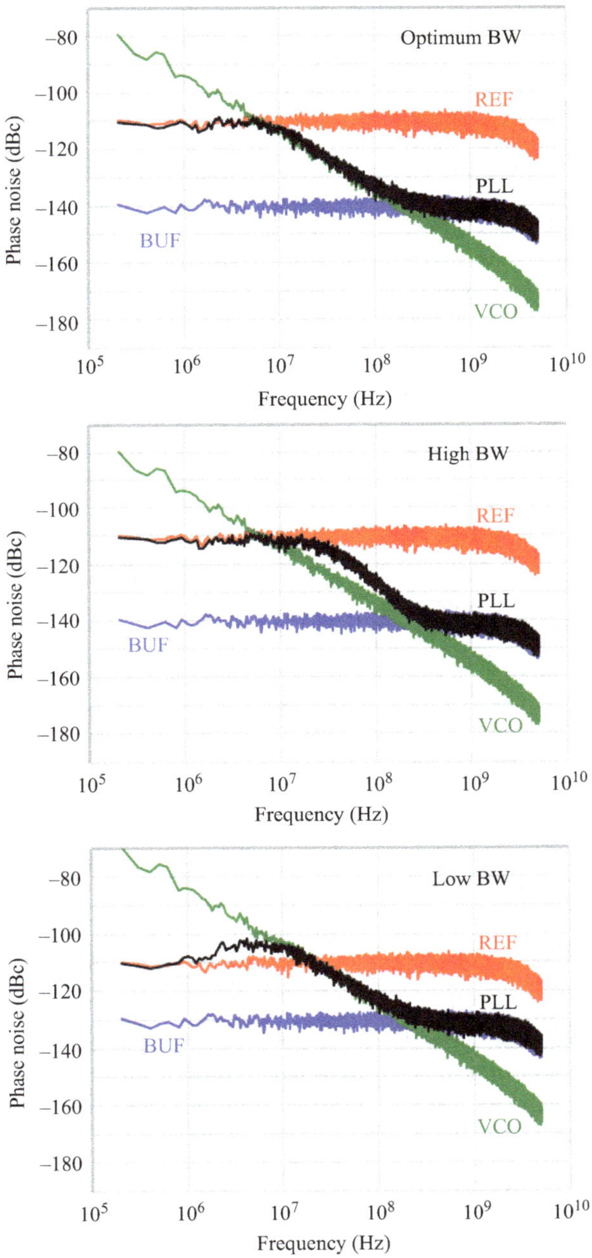

Figure C.1 Simulated phase noise profiles from the System Verilog model

```
real p_clk_jtt_in_flt_d = 0.0; // For implementing filter
real p_clk_jtt_in_d = 0.0; // For implementing filter
real p_clk_jtt_sum = 10*p_clk_ideal; // PM+FM jitter clock period
real p_clk_jtt_sum_d = 0.0; // For implementing filter
real p_clk_jtt_pll = 0.0; // PLL jitter clock period
real p_clk_jtt_rand_norm = 1.0; // Normalized PM jitter
real p_clk_jtt_in_norm = 1.0; // Normalized input jitter
real p_clk_jtt_acc_norm = 1.0; // Normalized VCO jitter
real p_clk_jtt_sum_norm = 1.0; // Normalized PMFM jitter
real p_clk_jtt_pll_norm = 1.0; // Normalized PLL jitter
real rand_jtt_mag = 1.0/2.0/3.14; // PM jitter magnitude
real in_jtt_mag = 1.0/2.0/3.14; // Input jitter magnitude
real acc_jtt_mag = 1.0/2.0/3.14; // VCO jitter magnitude
// Ideal clock generation
initial clk_ideal = 1'b0;
always #(p_clk_ideal/2) clk_ideal = ~clk_ideal;
// Jitter clock
integer my_seed = 567;
integer my_seed_acc = 1;
initial clk_jtt_PM = 1'b0;
initial clk_jtt_FM = 1'b0;
initial clk_jtt_PMFM = 1'b0;
// Jitter generation from buffer and VCO
always @(clk_ideal)
begin
// Mag of TIE by integrating white PN, 5.3 is a fitting param
rand_jtt_mag <=5.3* 1.0/2/3.1416*((f_clk*1e9)*(10.0**(white_pn/10.0)))
**0.5;
// Clock period with white jitter
p_clk_jtt_rand <= 1.0*p_clk_ideal
+ sine_jtt_mag*p_clk_ideal*sin(2*3.1416*sine_jtt_frq*$realtime)
+ 1.0*rand_jtt_mag*p_clk_ideal*1.0*($dist_normal(my_seed,0.0,1.0));
// Normalization for PSD analysis
p_clk_jtt_rand_norm <= p_clk_jtt_rand/p_clk_ideal;
// Mag of period jitter of VCO, 3.1416 is a fitting param
acc_jtt_mag <= 10/3.1416*white_cutoff/(f_clk*1e9)**0.5*(10.0**
(white_pn/20.0));
// TIE of accumulation jitter, accumulating period jitter
p_clk_jtt_acc <= p_clk_jtt_acc
+ 1.0*acc_jtt_mag*p_clk_ideal*1.0*($dist_normal(my_seed_acc,0.0,1.0));
// Normalization for PSD analysis
p_clk_jtt_acc_norm <= p_clk_jtt_acc/p_clk_ideal;
// TIE of accumulation jitter + white jitter
p_clk_jtt_sum <= p_clk_jtt_acc
+ sine_jtt_mag*p_clk_ideal*sin(2*3.1416*sine_jtt_frq*$realtime)
```

```
+ 1.0*rand_jtt_mag*p_clk_ideal*1.0*($dist_normal(my_seed,0.0,1.0));
// Normalization for PSD analysis
p_clk_jtt_sum_norm <= p_clk_jtt_sum/p_clk_ideal;
// Generating jitter clocks
if(clk_ideal==1'b1) clk_jtt_PM <= #(p_clk_jtt_rand/2) 1'b0;
else clk_jtt_PM <= #(p_clk_jtt_rand/2) 1'b1;
if(clk_ideal==1'b1) clk_jtt_FM <= #(p_clk_jtt_acc/2) 1'b0;
else clk_jtt_FM <= #(p_clk_jtt_acc/2) 1'b1;
if(clk_ideal==1'b1) clk_jtt_PMFM <= #(p_clk_jtt_sum/2) 1'b0;
else clk_jtt_PMFM <= #(p_clk_jtt_sum/2) 1'b1;
end
// Input jitter
always @(clk_ideal)
begin
// Mag of TIE by integrating input PN, 5.3 is a fitting param
in_jtt_mag <= 5.3*1.0/2/3.1416*((f_clk*1e9)*(10.0**(ref_pn/10.0)))**0.5;
// Clock period with input jitter
p_clk_jtt_in <= 1.0*p_clk_ideal
+ 1.0*in_jtt_mag*p_clk_ideal*1.0*($dist_normal(my_seed,0.0,1.0));
// Normalization for PSD analysis
p_clk_jtt_in_norm <= p_clk_jtt_in/p_clk_ideal;
end
// PLL jitter
always @(clk_ideal)
begin
// PLL BW translation for digital filter
par_pll_bw <= 1.414 * p_clk_ideal/(p_clk_ideal + 1.0/3.1416/pll_bw*1e9);
// High-pass filter of jitter generation
p_clk_jtt_sum_d <= p_clk_jtt_sum;
p_clk_jtt_acc_flt_1st <= (p_clk_jtt_sum - p_clk_jtt_sum_d) * (1.0 - par_pll_bw) + (p_clk_jtt_acc_flt_1st) * (1.0 - par_pll_bw);
// 2nd order
p_clk_jtt_acc_flt_d <= p_clk_jtt_acc_flt_1st;
p_clk_jtt_acc_flt <= (p_clk_jtt_acc_flt_1st - p_clk_jtt_acc_flt_d) * (1.0 - par_pll_bw) + (p_clk_jtt_acc_flt) * (1.0 - par_pll_bw);
// Low-pass filter of input jitter
p_clk_jtt_in_d <= p_clk_jtt_in;
p_clk_jtt_in_flt_1st <= p_clk_jtt_in_d * par_pll_bw + (p_clk_jtt_in_flt_1st) * (1.0 - par_pll_bw);
// 2nd order
p_clk_jtt_in_flt_d <= p_clk_jtt_in_flt_1st;
p_clk_jtt_in_flt <= p_clk_jtt_in_flt_d * par_pll_bw + (p_clk_jtt_in_flt) * (1.0 - par_pll_bw);
// PLL jitter
p_clk_jtt_pll <= p_clk_jtt_acc_flt + p_clk_jtt_in_flt;
```

```
// normalization for PSD analysis
p_clk_jtt_pll_norm <= p_clk_jtt_pll/(p_clk_ideal*1.0);
// Generating jitter clock
if(clk_ideal==1'b1) clk_jtt_PLL <= #(p_clk_jtt_pll/2+p_clk_ideal*1.0)
1'b0;
else clk_jtt_PLL <= #(p_clk_jtt_pll/2+p_clk_ideal*1.0) 1'b1;
end
endmodule : clkgen
```

Reference

[1] Drakhlis B. Calculate oscillator jitter by using phase-noise analysis. *Microwaves & RF*. 2001;40(1):82.

Appendix D
Noise sources in MOSFET transistor

In this appendix, noise sources in CMOS transistors, that is, thermal noise, $1/f$ noise, and shot noise, are briefly reviewed [1–8]. Generally, thermal noise and $1/f$ noise dominate the noise of MOSFET devices in normal operation. On the other hand, shot noise does not occur in an ohmic conductor so only CMOS transistors in subthreshold region exhibit the shot noise.

The thermal noise is caused by the thermal fluctuation of charge density in a conductor. The thermal noise has a uniform power spectral density regardless of the frequency and is independent of the conductor material. The thermal noise is modeled with either a resistance in series with a noise voltage generator or a resistance in parallel with a noise current generator as shown in Figure D.1. According to the Nyquist theory, the PSD of the voltage noise is given as

$$\frac{\overline{v_n^2}}{\Delta f} = 4kTR \tag{D.1}$$

where k, T, and R represent the Boltzmann constant, the absolute temperature, and the resistance, respectively. Equivalently, the current noise is obtained as

$$\frac{\overline{i_n^2}}{\Delta f} = \frac{4kT}{R} \tag{D.2}$$

Based on (D.1) and by integrating the resistance of a channel segment over the channel [6], the thermal noise of a MOSFET device is derived as

$$\frac{\overline{i_n^2}}{\Delta f} = 4kT\gamma g_m \tag{D.3}$$

where $\gamma = 2/3$ for a long-channel transistor, which comes out from the integration. However, it is hard to get a closed-form expression with an analytic approach for a short-channel transistor. Instead, experimental approach finds that γ increases to $1 < \gamma < 3$ in short-channel transistors, because of velocity saturation [5,9].

A beginning designer might be exposed to the word "kT/C" noise, which might confuse him/her that a capacitor can be a source of noise. Note that a capacitor is

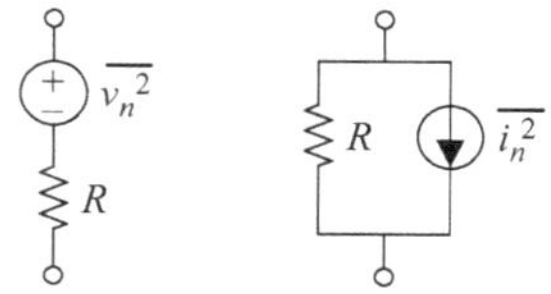

Figure D.1 Resistance with voltage noise and current noise

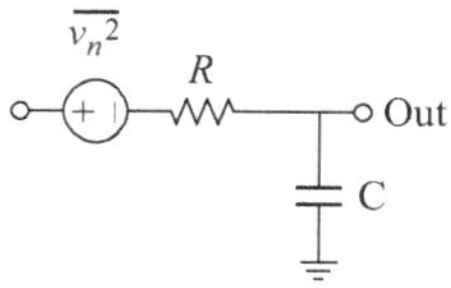

Figure D.2 Noise of RC circuit

not a source of noise, but a resistor is the source of the kT/C noise. Let us assume that we have a resistor which is connected to a capacitor as shown in Figure D.2. The resistor generates a voltage noise whose PSD equals 4 kTR up to infinite frequency. However, due to the RC low-pass filter, the high-frequency noise is filtered out at the output node. Since the cut-off bandwidth is 1/RC, the total integrated noise is obtained with

$$\overline{v_{n,\text{out}}^2} = \int_0^\infty 4kTR \cdot \frac{1}{1+(2\pi fRC)^2} df = 4kTR \cdot \frac{1}{2\pi RC} \cdot \arctan(\omega RC)\Big|_0^\infty = \frac{kT}{C} \tag{D.4}$$

Note that the integrated noise is independent of R, although the source noise is proportional to R. Intuitively, the noise power is proportional to R, but the bandwidth is inversely proportional to R, and therefore their effects cancel out.

The shot noise is caused when the number of electrons is not enough so that the current flowing is like pulse train by individual electrons randomly arriving rather than continuous flowing. The shot noise also exhibits a uniform PSD up to very high frequency (determined by the pulse-width of the current pulse by a single electron), and the PSD is given as

$$\frac{\overline{i_n^2}}{\Delta f} = 2qI_{\text{DC}} \tag{D.5}$$

where q is the charge of an electron (1.6×10^{-19}C) and I_{DC} is DC current. Detailed derivation of the shot noise is given in [2,5]. In contrast to the thermal noise, the shot noise is independent of temperature.

$1/f$ noise refers to a noise whose PSD increases as the noise frequency decreases, and therefore it is the dominant source of noise at low frequencies. The $1/f$ noise is observed in all practical active devices and electronic materials [10,11]. However, a fully satisfactory physical understanding of the $1/f$ noise has not been entirely established yet. For MOSFET devices, the number of carriers fluctuation theory [10] explains that the $1/f$ noise is generated by fluctuations in the number of carriers attributed to the trapping and detrapping of charges in traps located in gate dielectric surface [11]. In this theory, the $1/f$ characteristic originates from the distribution of the trapping time constant and summation of those distributed time constant. In addition, the magnitude of the $1/f$ noise is proportional to σ_N/N, where N and σ_N are the average number of traps and the standard deviation of number of traps. Since σ_N is proportional to $\sqrt{N}$, we can obtain

$$\frac{\overline{i_n^2}}{\Delta f} \propto \left(\frac{\sigma_N}{N}\right)^2 \cdot \frac{1}{f} \propto \frac{1}{N} \cdot \frac{1}{f} = \frac{1}{D_N WL} \cdot \frac{1}{f} \tag{D.6}$$

where D_N, W, and L represent average trap density, width of transistor, and length of transistor. We can observe that the $1/f$ noise is also inversely proportional to the dimension of a transistor, which means that it has become severer as the CMOS process technology scales down.

References

[1] Spieler H. *Semiconductor Detector Systems*. New York: Oxford University Press; 2005.

[2] Cristofolini L. Shot and thermal noise. Università di Parma, http://www.fis.unipr.it/~gigi/dida/strumentazione/harvard_noise.pdf.

[3] Instruments T. Noise analysis in operational amplifier circuits. Application Report, SLVA043B. 2007.

[4] Lundberg KH. Noise sources in bulk CMOS. Unpublished paper. 2002;3:28.

[5] McNeill JA. Noise in short channel MOSFETs. In *2009 IEEE Custom Integrated Circuits Conference*. Rome: IEEE; 2009 (pp. 567–572).

[6] Jordan AG, Jordan NA. Theory of noise in metal oxide semiconductor devices. *IEEE Transactions on Electron Devices*. 1965;12(3):148–156.

[7] Sarpeshkar R, Delbruck T, Mead CA. White noise in MOS transistors and resistors. *IEEE Circuits and Devices Magazine*. 1993;9(6):23–29.

[8] Nyquist H. Thermal agitation of electric charge in conductors. *Physical Review*. 1928;32(1):110.

[9] Scholten AJ, Tromp HJ, Tiemeijer LF, *et al.* Accurate thermal noise model for deep-submicron CMOS. *In International Electron Devices Meeting 1999. Technical Digest (Cat. No. 99CH36318)*. Washington, DC: IEEE; 1999 (pp. 155–158).

[10] McWhorter AL. *Semiconductor Surface Physics*. Philadelphia, PA: University of Pennsylvania Press; 1957. pp. 207–228.

[11] Zhang T. A statistical flicker noise analytical model in scaled bulk MOSFETs. Electrical Engineering and Computer Sciences, University of California at Berkeley UCBIEECS-2009-18. 2009.

Index

www.ingramcontent.com/pod-product-compliance
Lightning Source LLC
LaVergne TN
LVHW011851200926
848484LV00002B/438